A House in Fez

SUZANNA CLARKE

Born in New Zealand, Suzanna Clarke grew up in several parts of Australia. In her twenties she lived in a Welsh commune, an Amsterdam squat and a Buddhist monastery in Nepal. Apart from forays into contemporary dance and academia, she has worked as a photo-journalist for more than two decades, contributing to Australian and international newspapers, magazines and books and holding several exhibitions. She is currently the arts editor of a major Australian newspaper.

SUZANNA CLARKE

A House in Fez

BUILDING A LIFE IN THE
ANCIENT HEART OF MOROCCO

EBURY
PRESS

II

Published in 2008 by Ebury Press, an imprint of Ebury Publishing
A Random House Group Company

First Published by Penguin Group (Australia) in 2007

The Random House Group Limited Reg. No. 954009

Addresses for companies within the Random House Group can be found at
www.randomhouse.co.uk

A CIP catalogue record for this book is available from the British Library

ISBN 9780091925222

To buy books by your favourite authors and register for offers visit
www.randomhouse.co.uk

The names of some people in the text have been changed to protect their privacy

**Penguin Random House is committed to a sustainable future for
our business, our readers and our planet. This book is made from
Forest Stewardship Council® certified paper.**

MIX
Paper from
responsible sources
FSC® C018179

Printed and bound in Great Britain by Clays Ltd, Elcograf S.p.A.

For Meg and Henry, my parents, who began the adventure.

And for Sandy, my love, who shares it still.

RIAD ZANY

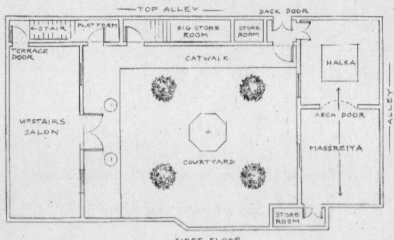

FIRST FLOOR

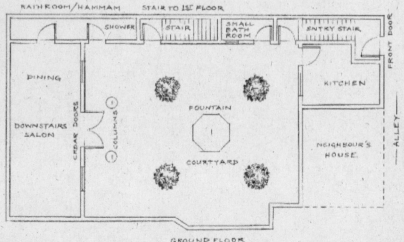

GROUND FLOOR

Prologue

aybe it was a fit of madness, but on just our second visit to the old Moroccan capital of Fez, my husband and I decided to buy a house there — as one does in a foreign country where you can't speak the language and have virtually nothing in common with the locals.

Morocco is only thirteen kilometres across the Strait of Gibraltar from Europe, but in almost every respect it might as well be on another planet. Situated in the north-west corner of Africa but separated from the rest of the continent by the vast Sahara, its Arabic name, al-Maghreb al-Aqsa, means 'the extreme west'. It is a land of cultural and ethnic amalgams, of Berber, African, Arab and, in more recent times, French and Spanish influences. The writer Paul Bowles called Morocco a place where travellers 'expect mystery, and they find it'. He wasn't wrong.

Our first visit, in 2002, had not been auspicious. I developed a raging case of the Moroccan quick-step after eating in a fancy palace restaurant, and like many tourists, we'd been fleeced in a carpet shop by a charming but wily salesman whose money-extracting technique had been handed down through centuries of dealing with naïve foreigners.

Nevertheless Sandy and I responded to Morocco in a way we had to no other country. We found it as multi-layered and intriguing as the patterns in the tile work adorning the buildings, each of which has its own hidden meaning. Morocco has the mystique of a land from the Old Testament yet appears to be coping comfortably with modernisation. Internet cafés rub shoulders with artisans' workshops; peasants on donkeys trot beneath billboards advertising the latest mobile phones. Outside mosques, running shoes are lined up next to pointy-toed babouches. In the souks, women wearing long robes and headscarves escort daughters with beautifully cut hair and high heels. You can eat at a street stall, in a Parisian-style café, or next to a tinkling fountain in an ornate courtyard. You can find yourself in the midst of a crazy, honking traffic jam, or dodging donkeys in cobbled alleyways, or riding a camel in the solitude of the Sahara.

Fez, in particular, won our hearts. Its location in a bowl-shaped valley protects it from the baking heat of summer and the bone-chilling winds that race down from the Atlas Mountains in winter. The ancient Medina, or walled city, is the cultural and spiritual core of Morocco and is at once a delight and a challenge, an assault on the senses. Its countless narrow alleyways are not a museum but a living community, abuzz with Fassis going about their business. The largest car-free urban area in the world, the Medina is home to quarter of a million people, but very few of them are expats. And unlike in Marrakesh or Tangier, tourists are more of a sideline than the main event.

Within the Medina's thick sandstone walls lie thousands of flat-topped houses. Viewed from a distance, they are impenetrably dense and complex, like a nest built by cubist termites – a room over a neighbour's stairs here, another room stretching over next-door's kitchen. The houses offer blind faces to the street; the only clue to what lies within are their intriguing doorways, decorated with intricately carved wood or with hinges showing the hand of Fatima, the Prophet's daughter, whose influence is believed to ward off the evil eye.

As is the case everywhere in the Medina, the dwellings of the rich sit next door to those of the poor, although all appear similarly drab from without. In Islamic philosophy, God and family are more important than showing your wealth to the outside world. Scattered among the houses are green-tiled minarets, from where five times a day a melodic voice echoes across the rooftops. Then another joins in, and another and another, until the whole city reverberates with the muezzins' calls to prayer.

There were aspects of Morocco we found confronting – the beggars, child labour, animals being treated as simply tools or pests. But the negative factors were part of a complex society, and were offset by the warmth of the people, the cultural and historical richness. It was a country where we both felt more alive than anywhere else, our every sense engaged.

Back at home in Brisbane after that first trip, our far too expensive Berber carpet – the colour of sunset on the sands of the Sahara – lay on the floor of our living room, reminding us of the sensual beauty of Morocco, and of our ignorance of the traps of that enormously different and distant culture. We found it hard to settle down. The advertising that bombarded us daily seemed more intense, the supermarkets soulless. Most of the food sold there was so far from its origins as to be unrecognisable, obscured by packaging designed to conceal the contents. Even the fruit and vegetables were plastic-wrapped, some of them coming from halfway around the world for me to buy, wilted, at many times the original price. And we in the 'developed' world were under the illusion that we had the best of everything.

Where were the lively, crowded markets that echoed the way humans had traded for centuries? Where was the produce piled high in the open, direct from the farm, allowed to ripen on the vine or on the tree or in the earth, and only available in season? In Moroccan markets you could smell, handle and often taste things before buying. You looked the farmer or the stallholder in the eye as you quibbled in a good-natured way about the quality and price.

On days when the walls seemed to be closing in, I would trawl the Internet, looking at sites on Morocco. Among them were a number offering real estate for sale, and slowly an idea started to form. Why not buy a house there? Living in Morocco for part of the year would enable us to explore its radically different reality in a much more profound way than just visiting. The notion seemed fantastical, but it grabbed my imagination and began to dominate my daydreams. I had a romantic vision of ancient walls touched by golden light, within which generations of lives had been lived. A house waiting for us to save it, to bring it back to what it once was — but perhaps with a better bathroom.

One of the reasons Sandy and I get on so well is that he never stops me from pursuing mad schemes; in fact he frequently gets more caught up in them than I do, usually about the time I'm getting cold feet. In the decade we've been together, his enthusiasm has got many of our projects over the line, and so it was in this case. Watching me become frustrated with high-priced houses in tourist enclaves, he said, 'Why don't you look for a house in Fez?'

It made sense as soon as he said it. The kind of place we were after was unlikely to be advertised on any website, but would be hidden deep in the Medina, where few foreigners lived.

There were obvious drawbacks, like the nuttiness of buying a house on the other side of the planet, a leg-cramping, bloodclot-inducing, 26-hour plane flight away. And just when would we actually get to spend time there? Our jobs consumed our lives — I worked on a busy metropolitan newspaper and Sandy had a national radio show, and the news does not conform to regular working hours. When exactly would we fit in a commitment to a property in another country?

Moreover we had absolutely no idea how to go about buying a house in Fez. There was no help on hand, no DIY manual, no *Morocco for Dummies*. And there was the problem of what to do with it when we weren't there. I had come across stories about vacant houses being ransacked, stripped of their doors and light fittings. One old house had collapsed after thieves removed the supporting beams.

As wage slaves with major mortgage commitments, we'd be moving way out of our comfort zone. And what about the language difficulties? In most of Morocco, with the exception of the major tourist areas, English is a rarity. I spoke a bit of schoolgirl French, and Sandy could muster a dozen or so words of Darija, the Moroccan dialect of Arabic.

The arguments against were endless, but the idea refused to go away. It hung around and fermented, bubbling in shadowy corners of our minds. What eventuated was to prove utterly different to what we'd imagined. Naïve and unprepared, we stepped into the unknown, and into the most intense and exhilarating experience of our lives.

When we mentioned our fantasy of buying a traditional Arab-style riad, or courtyard house, in Fez to a friend he said dismissively, 'What a terribly nineteenth-century thing to do.'

He had a point. For most of my life I have been enchanted by tales of early European women travellers, such as Isabelle Eberhardt and Jane Digby, who broke out of the strictures of lives far more confining than my own and found another way to live, in Arab cultures. Of course, such adventures are only romantic if you ignore the fact that Eberhardt, who disguised herself as a man, contracted syphilis then drowned penniless and alone, washed away with her final manuscript.

There were other acquaintances who, post-September 11, asked, 'Why would you want to buy a house in a Muslim country? They hate us.'

This was easier to counter; we knew it simply wasn't true of

Moroccans, who can be friendly and hospitable to the point of overwhelming. We were also aware that people in Western countries tend to view Muslim nations as a monolithic bloc, whereas there are many cultural differences between them, despite common elements. The present King of Morocco, Mohammed VI, has built bridges with the West, and was the first Muslim ruler to express sympathy for the United States following the destruction of the World Trade Center.

Besides, we saw our venture as an opportunity to explore Islamic culture further, and to gain a deeper insight into why the Way of the Prophet has thrived for so long.

There was a certain inevitability about my interest in Morocco. My parents had visited the country in 1961, long before it was fashionable to do so. At that time only a handful of hardy souls, forerunners of the hippies who were later to invade, made their way down from Spain, following the sun. Enthusiastic young travellers, Meg and Henry drove their Volkswagen beetle around Morocco's few, mostly unpaved, roads, and after some hair-raising adventures ended up pitching their tent in the camping ground at Marrakesh. It appears they had a particularly convivial time there, because I was born exactly nine months later.

Being conceived in Morocco and growing up in New Zealand, I learned to walk at a shuffle in my father's Moroccan babouches, surrounded by mementos from their visit. One of their more colourful tales was of the night they camped on the side of a road high in the Atlas Mountains. In the middle of the night, a truck came winding up from the valley below, its headlights swinging across my parents' tent. They stayed in their sleeping-bags, hoping it would continue past, but when the truck drew level with their car its engine stopped.

Rocks crunched as footsteps moved towards them. My father, deciding drastic action was called for, unzipped his sleeping-bag, grabbed the tomahawk, and when he judged the moment was right, leapt out wearing only striped pyjama bottoms and swinging the tomahawk above his head, bellowing a Maori haka. '*Ka mate, ka mate, ka ora, ka ora, Tenei te tangata puhuruhuru . . .*'(It is death! It is death! It is life! It is life! This is the hairy man . . .')

He must have been a terrifying sight, because four djellaba-clad men ran to their truck and sped off into the night. I wonder if those men now tell their grandchildren about the time they stopped to help the occupants of a car they thought had broken down and were confronted by a screaming madman.

By early 2003, the idea of buying a house in Fez was sufficiently implanted for Sandy and I to start putting money aside. Forget buying a new car, clothes, or even basic house maintenance – this was our escape fund. The second part of our plan was to get ourselves back there without dipping into it.

The previous year, we had taken holiday jobs as tour managers for a small group of well-heeled tourists in France and the UK. Now we proposed Morocco as a destination. After studying and planning for months, we flew out in October and showed our group as much of Morocco as was possible in three short weeks, with the help of a local guide. It was spring and the countryside had transformed from the dry brown of our previous trip to lush green, liberally sprinkled with poppies, irises, daisies and other wildflowers.

Tour leading entailed a level of luxury entirely different from what we were accustomed to, and it wasn't always for the better. Inside a four- or five-star hotel, with the exception of a few designer touches, you could be almost anywhere in the world. There was a sameness, a monotone, that irked us both, but for some of our guests such accommodation still wasn't up to par.

'I thought there'd be hairdryers in the hotel bathrooms,' one wealthy woman whined.

Many people in Morocco don't even have running water, never mind washing machines or refrigeration, so hairdryers are not considered an essential item. But to her mind they were, and she left us in no doubt that Sandy and I were negligent for not ensuring their presence.

We had gone to great lengths to give our clients a varied cultural experience, but it wasn't always to their taste. We arranged some of our city stays in beautiful riads, which, being several hundred years old, had rooms that were not standard sizes. Nor were the bathrooms always ideal.

'Why can't we stay in the new part of town?' an elderly man complained. 'I'm tired of all this history.'

It wasn't a sentiment Sandy and I shared. When we reached Fez we stood gazing down over the ancient walls of the Medina and the decrepit houses within, just biding their time for people with the vision, money, time and energy to restore them. We confirmed to one another that those people should include us.

During my Internet surfing, I had come across the website of an American living in Fez who claimed that, although the Medina is the best-preserved mediaeval walled city in the world, its architectural heritage is under threat. Many Moroccans cannot afford to maintain the houses they live in, let alone restore them to their original splendour. If those with the means to do so, foreigners included, were to rescue some of the significant houses, this would make a big difference to the preservation of the Medina. The key point, the website argued, was the need for proper restoration, as opposed to modernisation. And rather than gentrification, a healthy mix of rich and poor living together, as had always been the case in the past, was the ideal.

One night, Sandy and I escaped from our tour group to meet the website's author, David Amster, over a drink. Originally from Chicago, David was in his mid-forties, amiable, intelligent, with a wonderfully dry sense of humour. His full-time job was Director of the American Language Center in Fez, and he had called the city home for seven years. He was a passionate advocate for the traditional architecture of the Medina, about which he was extremely knowledgeable, giving lectures to visitors from the Smithsonian Institute once or twice a year. He owned five houses in Fez, one of which was a ruin. It had completely collapsed, leaving only columns and mosaics, which made it resemble a Roman temple.

'What are you going to do with it?' Sandy asked.

'At the moment,' David said, 'I have no idea. The site's under six feet of rubble. It'll take hundreds of donkey hours to move.'

But he didn't appear too concerned. 'Look,' he continued,

pulling out of his bag a sliver of wrought iron that looked like an oversized thumb tack and passing it to me. 'Isn't it beautiful? It's a handmade nail.'

This nail, several hundred years old, was to be the prototype for those David was having made. After months of searching, he had finally found a man with the necessary skills, but there was to be one small difference: the heads of the new nails would be slightly bigger, so that future experts could tell old from new.

Sandy and I exchanged glances. This was restoration on a whole new level. It wasn't just popping down to the hardware store and getting a bag of cheap Chinese nails and a bit of four-by-two. There was obviously more to doing up a house here than we realised.

'Would you like to see the house I live in?' David asked. 'I've had a builder and his team working on it for four years.'

He didn't need to ask twice. Leaving the modern quarter, we passed under the Bab Bou Jeloud – the Blue Gate – and into the Medina, then down innumerable dark alleys populated by cats who blinked at us, or slunk away on our approach. In the moonlight, one alley looked much like another and I was amazed at David's unerring sense of direction. We finally arrived at a large doorway, and as he ushered us in the first thing I noticed was the strong, fragrant smell of freshly sanded cedarwood.

We were standing in a dimly lit courtyard with an atrium stretching high above. All around was a wealth of decoration: blue, green and white tilework – or *zellij* – intricate plaster and wood carving, and two enormous cedar doors. It was like a jewel box, its beauty so overwhelming it was hard to imagine living in such a palace.

In contrast, the furniture was surprisingly Spartan: a bed, a single chair, nowhere to cook as yet. But who needs possessions and comforts when you wake up every day to such splendour? We had never envisaged anything like this. The prospect of restoring such a house was at once daunting and thrilling.

A few days later, I farewelled Sandy and the tour party at Casablanca airport. Sandy had to return to Australia for work, but I caught the train back to Fez, with the intention of finding us a house.

ez was once the largest city on the planet. Founded in 789, it became the centre of Moroccan scientific and religious learning, a status due to the altruism of a remarkable woman named Fatima al-Fihria. One of a group of refugees who fled religious persecution in Kairouan, Tunisia, in the ninth century, Fatima was from a wealthy merchant family and used her inheritance to start a place of learning. Karaouiyine University was completed in 859 and is the oldest educational institution in the world. Classes in religion are still held at the complex, which also contains a mosque and a library.

Fatima's act was even more altruistic than it might appear. Being a woman, she couldn't actually attend the university herself, but plenty of men did – Muslims and Christians from all over North Africa, the Middle East and Europe. In fact Karaouiyine had a major impact on mediaeval Europe. In the tenth century, Arabic numerals, including the concept of zero, were taken back to France by a student who went on to become Pope Sylvester II. He used his newfound understanding to invent a more efficient abacus, the basis of modern computing. Karaouiyine University also rejuvenated and spread the Indian concept of the decimal point, for which accountants are no doubt eternally grateful.

Modern-day Fez has three distinct sections. The Medina is the oldest and is known as Fez-al-Bali, meaning Old Fez. The second, Fez Jedid, or New Fez, lies uphill from the Medina and dates from 1276. It includes the Mellah, the old Jewish quarter. The third section, the Ville Nouvelle, is the administrative and commercial centre. Having learned from their mistakes in Algeria, the twentieth-century French colonisers resisted bulldozing the Fez Medina in order to modernise because the locals tended to get upset, with nasty consequences. Instead they situated the Ville Nouvelle several kilometres away. With its broad avenues and street cafés, it looks as if it were modelled on Haussmann's Paris.

As my school French had not equipped me to do much more than catch a train and order a coffee, and my Darija being limited to 'please' and 'thank you', I hired a translator, a polite young man named Nabil. He seemed bemused by my desire to live in a city that he and his friends would have left in a flash, if given a work visa in a Western country.

David had written me a list of houses he considered architecturally significant and which were in our price range. In Fez there are two types of houses, dars and riads. While both are centred around courtyards, a riad is generally much larger, its distinctive feature being a garden, or at least a lemon and an orange tree. Riads are rare in Fez and usually expensive.

'You're a riad person,' David told me one day, after I expressed a desire for some light and greenery. The problem was we were on a dar budget.

As my search proceeded, Nabil's back, framed by the towering walls that enclose the Medina's alleyways, became a familiar sight. During the day, the alleys were a constantly changing panoply of passers-by, children playing, salesmen touting, donkeys carrying goods. House doors were often open and old people sat on their steps warming themselves in the sun. At night, though, as I soon discovered, the alleys and their entranceways became a haven for teenagers' secret rituals of hashish and stolen kisses.

Following Nabil around, I began to understand that the Medina was a place where helping your neighbours was not an option but a necessity. Once, Nabil paused at a doorway where a veiled girl of about thirteen stood holding a tray of unbaked bread. He asked her a question, she pointed in reply, and without hesitation he took the tray and continued walking.

'Do you know her?' I asked him.

'No, but anyone who lives in the Medina and is walking past a bakery will take someone's bread.'

Neighbourhood bakeries, where families can send not only their bread but biscuits, cakes, even whole legs of lamb or a pizza, are one of the five essential facilities that Fez's thirty-odd quarters have in common. According to Islamic tradition, the others are a mosque, a school, a communal bathhouse and a fountain.

'How does the baker know whose bread is whose?' I asked.

He looked at me as if I were lacking intelligence. 'Every family makes their bread slightly differently, and the baker will have been baking it for many years, so he just knows.'

Coming from a country where the staples of life are mass-produced, I found this unaccountably wonderful. 'He just knows,' I muttered under my breath as we continued on.

Although at first the Medina seemed a confusing maze, I soon discovered that it is laid out in an organic and logical fashion. Like blood vessels leading into veins, small alleys where people live join streets with tiny shops that sell everyday necessities. In turn, these streets lead to main roads where the souks, or markets, are, and these roads eventually connect with the city's two major arteries – Tala'a Kbira (the big rise) and Tala'a Sghira (the small rise).

Running beneath all these streets and alleys is a complex network of water channels which supply households, public fountains, and industries such as the tanneries. Part of the construction of this ingenious water system, which dates from the eleventh and twelfth centuries, involved dividing the Oued Fez, the river that runs through the Medina, into two branches – one supplying clean water and the other carrying away effluent. But by the 1960s the system was faltering: the flow of the Oued Fez was reduced because of building in the catchment area to house rural refugees, driven to the city by a drought lasting more than a decade. A dam was built, and nowadays many of Fez's public fountains no longer flow as they once did.

Fez was not built to a master plan, but instead reflects the

Islamic principle of social order, with individual expression sub-
sumed into a harmonious whole. For this reason it is regarded
as the model Muslim city. Its streets are not merely visually stimul-
ating, but give the other senses a workout as well. The spicy scent
of tagines wafts out of kitchen windows, mingling with the yeasty
aroma of bread and cakes from communal ovens. There's the
heady fragrance of fresh cedar being shaped into a door, a window
frame or a piece of furniture. Less attractive is the acrid stench of
glues and solvents. Donkey dung litters the wider alleys, and I
soon learned to watch my step. Just as quickly I became inured
to the reek of cat and human pee.

Being a female turned out to be an unexpected advantage in
my house search. Usually only women are at home during the
daytime, and they're generally reluctant to admit a strange male
to the house. But when they saw me they would visibly relax, and
we were always invited inside. In the space of a few days I saw
so many houses that they began to meld into each another, and
it was difficult to remember which kitchen or terrace went with
which house. To overcome this, I took photos and notes and went
over them at the end of the day.

David had given me some tips for determining the age of a
house. He told me that those with *masharabbia* — intricate wooden
mesh screens on upstairs balconies and on upper windows facing the courtyard — were built before the late nineteenth century.
Under Islam as practised then, the faces of women could not be

seen by men outside their own families. This meant that when strangers came to the house the women would retreat to sit behind these screens, where they could see what was going on in the courtyard while remaining hidden from view. As attitudes became more relaxed during the nineteenth century, *masharabbia* began to be replaced with less expensive wrought iron.

With other information David gave me, I was soon able to estimate the age of a house from the height at which tiles stopped on the walls, and from the colours and decorations used. But it wasn't always straightforward, as plasterwork and *zellij* were usually renewed every hundred years or so. A house could be much older than it appeared on the surface.

I saw some once magnificent but now forlorn houses that were crying out for restoration, deserted by wealthy families who'd left for Rabat when the French protectorate made that city the capital instead of Fez. Now these huge crumbling edifices were full to bursting with squatters, drought refugees from the countryside. Where one family might previously have lived, now there were five or six. Having had no maintenance for decades, their once grand rooms had an air of neglect and quiet desperation.

'What happens to the people here if the house is sold?' I asked the agent as I inspected one particularly glorious but decaying example, the one-time residence of a Sufi saint. Every room was occupied by a different family, and the outstanding woodwork was cracking with age and exposure to the weather.

The estate agent had also been recommended to me by David. Larbi was a small, grey-haired man who refrained from the incessant sales patter his Australian counterparts engaged in. Maybe this was because he spoke not one word of English.

'They are paid to move as part of the deal,' he told me via Nabil.

That made me feel less guilty, but we couldn't afford to do the amount of work required to save the Sufi's riad, even if we'd been able to scrape together the asking price. I just hoped someone with the money and the know-how would come along soon.

Some houses I saw were too far gone to consider, their structural problems severe enough to overwhelm even the most ardent restorer. There were fissures in walls and undulations in floors that spoke of subsidence over the centuries, poor foundations, and other, unknown forces at work in the earth beneath.

My aim was to find a house that had been spared the process of modernisation. In poor communities, the challenges of the present often take precedence over saving the past, and the Fez Medina is no exception. When Moroccans come into money, most want what people the world over desire – modern amenities. Things that look shiny and new, just like on television. This can spell disaster for an ancient house.

David had taken me to one riad where the traditional *zellij* had been covered over with shiny grey Chinese bathroom tiles. The walls had been painted an institutional green, and the wall fountain, which would have once been beautiful, had been ripped out and replaced with a laminated hand basin illuminated by a fluorescent light. The combination of the décor and high walls made me feel like I'd been transported to an episode of *Porridge*, the British television comedy about life in prison.

Another riad David showed me was important enough and old enough to be listed on a register of significant buildings. Inside, the first thing that struck me was the smell — an overwhelming odour of stale urine. I had to ignore the objections of my nose as David excitedly pointed out the exquisite wooden carving around the atrium, and the fine *zellij* work on the wall fountain. It was obvious that at one time this house had been really cared for.

The riad had two large salons facing one another across a courtyard, and a sitting niche on another side that had been enclosed. Beside the fountain, stairs led to a *massreiya* — a highly decorated apartment usually reserved for the eldest son or male visitors. Up in this *massreiya*, a woman was nursing a baby while a pot of chickpeas bubbled on a gas burner beside her. She smiled at us, unconcerned by our intrusion.

The riad's small appearance was deceptive, belying the rabbit warren of rooms we discovered. In two of the larger rooms, ancient and beautiful tile work was covered with manure. Evidently these were the farm rooms, and I felt a pang of sympathy for the animals forced to live there.

Up on the terrace was a studio, and a makeshift and very rustic kitchen. David was in his element, intrigued by the details of how this chimney was constructed and by that painting on the ceiling of the studio. But looking out from the rooftop, we discovered the source of the overpowering stench. The house overlooked one of Fez's most famous tourist attractions — the thousand-year-old Chouwara tanneries. Essential ingredients in the bleaching of the hides are pigeon poo and goat's urine. Next to the bleaching pits are dozens of vats filled with blue, red, yellow and brown dyes, in which leather workers spend their days immersed thigh-deep. Colourful and interesting, but hardly an ideal neighbour.

David tried to convince me to buy the riad. 'It's a real gem,' he enthused.

There was no doubt it was an extraordinary piece of architecture, but there was the small problem of the smell.

'What you need to find,' I told him, 'is a rich person with no olfactory nerves. And preferably a loner, because no one will ever come to visit.' It wasn't me.

Despite the frustrations of house hunting, I loved staying in the Medina. The place was alive with sounds from morning to night. During the day, there were cries of '*Balak, balak!*' – 'Take care, take care!' – as mules and donkeys, laden with produce, trotted through the alleys. Occasionally I would hear the sound of African drumming, amplified by the high walls, and snatches too of Arabic music, and always the shouts and squeals of children playing. Walking at night, I was never far from the sound of water, gurgling below the paving stones or tinkling from some unseen fountain. Sometimes there were high-pitched ululations from far over the rooftops as a wedding celebration continued into the small hours.

If I woke before dawn I would hear a melodic, lyrical song coming from the nearest mosque. This was to comfort insomniacs and the sick, and was followed by the muezzin's first call to prayer. The cry, 'It is better to pray than sleep' spread from mosque to mosque, consuming the city in a chant that echoed across the rooftops and rang around the courtyards, luring the devout from their beds. It gave the Medina a sense of otherworldliness, of another layer hidden below the surface of everyday life.

After about a week of looking at several houses a day, I saw two I liked in succession. The first was a larger dar with pristine blue and white *zellij*. Although the tiles were new, they had been skilfully laid in the old style, and I was not surprised to learn that the house had been the home of a master *zellij* craftsman. The upper rooms had cedar shutters and window frames that were in excellent condition, but the grand doors of the salons were missing.

Larbi claimed these could easily be replaced, but I wasn't so certain. I had seen one set for sale in Marrakesh for the equivalent of fifty thousand Australian dollars. This was more than the price of the house, and two sets were needed. It also occurred to me that replacing them would no doubt entail removing the doors from some other old house, so all we'd be doing was passing the problem on.

The second house was a riad, centrally located off the main tourist thoroughfare, the Tala'a Sghira, at the end of the narrowest alley I'd ever seen — only slightly wider than my shoulders. I wondered how they managed to get furniture into the house. Through the front door, I followed a long passageway and emerged via another doorway into a small but attractive courtyard with two trees — a straggly loquat and a lemon — and a beautiful wall fountain. Upstairs, above the ground-floor salon, was an enormous room the size of a European apartment, with stained-glass windows and reasonable *zellij*. There were several other small rooms on the same level, and from the roof a view over the Medina. Not a spectacular view, but taken as a whole the property was the best I had seen, and it was in our price bracket. It didn't scream to me, 'I'm the one, buy me,' but my week was up and it was either this house or wait another year.

Back in Brisbane, I showed Sandy the photographs and described the riad. He was much more enthusiastic than I, but maybe I was housed-out. We decided to go ahead and buy it, but there was one major problem – the vendors spoke only Darija. As Australia wasn't exactly flush with Darija-speakers, it was necessary to get someone in Morocco to place our offer, and the most obvious choice was David. I was reluctant to impose further on his goodwill, since he'd been so generous with his help already and had an extremely busy job, but there didn't seem to be an alternative. I emailed to ask if he would act as our agent, with us paying him for his services. He said yes, but did not wish to charge anything.

Over the next several months, as negotiations became increasingly complex, I wondered if David ever regretted his generosity. There was a bit of argy-bargy about the price, but as usual both sides compromised and we ended up somewhere in the middle. There was an engineer's report to organise, and paying the deposit turned into a major drama when the money disappeared between banks, only to turn up weeks later, just as we were getting desperate. No doubt a bank employee somewhere had been making the most of the currency markets.

In preparation for life in Morocco, Sandy and I took up French lessons with a young Belgian living in Brisbane, who obligingly doubled as a translator when we needed to ring the notary in Fez. With a growing sense of excitement we arranged for settlement the following January, when we planned to return. That was eight months away, but seemed more than adequate even for the most convoluted of bureaucracies.

But as the months dragged on, it turned out that the vendor was missing a vital piece of documentation – the signed transfer from the previous owner – without which he was legally unable to sell the riad. Since that owner was now dead, the vendor was going to try to get the man's numerous relatives to sign a new version of the document, and he claimed everything would be sorted out by the time we arrived, *inshallah*.

Now, *inshallah*, meaning 'God willing', is a wonderfully useful expression. It means that things will happen in their own time, if God wills it. If I say to you I will do something, *inshallah*, it means that I have every intention of doing so, and if I'm prevented it's not my fault but is Allah's will. In Morocco, it's no use getting impatient and frustrated by delays, or blaming people. Things simply are. So it was with the transfer document. It would appear, we kept being told, by the settlement date, *inshallah*.

Obviously Allah didn't will it, because when January came the document still hadn't materialised. Sandy had gone to Morocco a week ahead of me, and one night he phoned to tell me the deal was off. There were about ten relatives involved, all of whom had been traced except one, and he hadn't been heard of in years. The only alternative was taking the matter through the courts, and there was no way of knowing how long that might drag on. So we were not going to be able to live in our riad after all. I was bitterly disappointed, and even more dismayed at the thought of having to begin the search all over again.

But Sandy was his usual optimistic self. 'Never mind,' he said, 'I have a plan B.' He had already found another house.

When I first saw Sandy's plan B it was in the company of some half a dozen people, including David, Larbi, Nabil, and Sandy's daughter Yvonne, who had just arrived from Ireland with her husband and two small children. Sandy had made an offer to the owners, contingent on my agreement, and everyone was waiting anxiously for my reaction. Sandy followed me around, pointing things out and saying, 'Isn't it wonderful?' while I tried to make an objective decision.

The new house was indeed a proper riad, situated at the end of an alley in one of the oldest parts of the Medina. You entered a carved wooden door, ascended some stairs in a corridor, and arrived in a lovely courtyard of about a hundred square metres, complete with an orange and a lemon tree and an attractive fountain in the centre. At one end of the courtyard was the kitchen, at the other a large salon, and on the right-hand side were doorways leading to two separate bathrooms and to a staircase to the floor above.

On the fourth side of the courtyard was a large and unusual feature wall, the central five metres of which was recessed by a metre. A buttress descended on the left-hand side of this recess, ending in a scallop shape, in the middle of which was a small spy window. We later discovered it had unique acoustic properties, allowing anyone sitting behind it to hear even the slightest whisper in the courtyard.

Two tiled columns stood in front of the downstairs salon, which had a set of massive decorated cedar doors. To either side of the doors were tall windows, framed with exquisite, hand-carved plaster. On the floor above this salon was another room of similar proportions, with its own set of beautiful doors.

Above the kitchen were two rooms large enough to be a self-contained *massreiya* apartment with its own entrance. The tops of these walls had a band of intricately carved plasterwork inset with rare and expensive coloured glass, known locally as Iraqi glass. And the first of these rooms contained the architectural treasure of the house – a huge ceiling with a spectacular radial design of carved and painted cedar.

'Museum quality,' David informed me. 'Truly wonderful.'

The house was indeed gorgeous, but I seemed alone in noticing that the beautiful ceiling was sagging on one side and the end section of the *massreiya* appeared to be in danger of falling down. There were a multitude of other repairs needed besides.

Anticipating my concern, Sandy had engaged the city engineer to inspect the house. What I managed to decipher of his report was amusing, but hardly encouraging.

Bending to the level of the floors of the room left lateral façade; detachment of the beams made of wood coupled supporting the catwalk; presence of beam of iron that encouraged the cracks of the wall and the lowering of the floor; presence of cracks deep to the level of the wall façade lateral internal left supporting the floor of the aforesaid room; rot of the tips floors of wood to the level of the hamman giving on the public way; flambement *to the level of the base wall and presence of humidity exaggerated caused by infiltrations of the public way toward the wall in question; detachment of the beams of the parallel walls to the wall of the giving façade on the alley; the soil of occupation of the patio is stuck out because of the roots of the trees that treaty would be necessary so that it reached the walls carriers of the building there. The physical state of the building is a little graduated and necessity of the funding works has know, including repairing the coming down of the pluvial waters.*

The '*flambement* to the level of the base wall' sounded like a real problem. I visualised the entire thing erupting in flames like a baked Alaska. It all sounded dangerous, and horribly expensive to fix. Nor was it clear to me how one made a treaty with the trees. Wall carriers were presumably the foundations, but 'repairing the coming down of the pluvial waters'? I guessed that meant mystified plumbers shaking their heads and doubling their prices.

I felt hesitant. This house needed far more work than the one we had failed to buy. True, it was lovely, but I was reluctant to take on a project of such magnitude. Had I been by myself, I think

I would have passed it up. Prior experience with renovations in Australia had taught me that whatever you think is needed is almost always an underestimation.

Sandy was hovering impatiently, waiting for my decision. I put him off, saying I would think about it. He was clearly disappointed, wanting me to share his enthusiasm, but my architect father had drummed into me from an early age the need to look carefully before committing.

The following morning, I went over the riad again with the engineer, deciphering his report, while Sandy stayed at the hotel nursing a bad case of the flu. The engineer, Salim, worked for the Agency for the Dedensification and Rehabilitation of the Fez Medina, but was more than happy to do a spot of well-paid moonlighting. He had a moustache that would have done George Harrison proud in his Sergeant Pepper days and he stroked it thoughtfully as he followed me from room to room.

'What about this?' I asked, pointing out the huge bow in the catwalk that joined the two upstairs wings, where two rotting beams were slowly parting company.

'No problem,' he replied breezily, and began drawing elaborate diagrams of how scaffolding could be placed to effect a repair. I tried to ignore the guillotine shape it seemed to resemble, although I did remember that a few months earlier a house had collapsed into a mosque, killing eleven people as they worshipped.

Salim's optimism was as rampant as Sandy's, and listening to either of them you would have thought that restoring a 300-and-something-year-old house was as easy as knocking up a garden shed. Still, compared to many of the houses in the Medina, this one was in prime condition. And it did have a terrace with one of the most spectacular views in Fez, all the way down the valley to the Atlas Mountains. The panorama was breathtaking: hills covered with cube-shaped houses to the left, sweeping down to Mount Zalagh. Here and there were the spires of minarets. In the far distance was a plume of black smoke from the potteries, which burned olive pits as fuel in the kilns. If you removed the satellite dishes, the view could have been straight out of the Old Testament.

When I returned to the hotel Sandy, through his sniffles, was still determined to be enthusiastic. 'The house is wonderful,' he said. 'It's a real find. Everything will be all right, you'll see.'

I decamped and went for a coffee with David, who didn't make things easier by pointing out that the riad was a significant piece of architecture and definitely worth saving. Eventually, bowing to the inevitable, I agreed to buy it, still harbouring considerable misgivings.

A couple of days later, we were ready to confirm the purchase. Sandy, rarely sick, had to struggle to get out of bed. He had an extremely high temperature and trouble standing. With the aid of medication, Nabil and I propped him upright long enough to get him into a taxi to the bank, to arrange our initial payment of fifty per cent of the price.

In Australia this would have taken all of ten minutes, but in Fez it was two complicated and frustrating hours before all the formalities were completed. The charming young bank executive

was so helpful that every time a new person came through the door
he would stop working on our account and switch his attention
to them. It took him more than an hour to fill out our one-page
form. This is how things are done in Morocco, I kept reminding
myself, taking deep breaths. It was just as well my high-school
French didn't run to swear words.

Like many people in the Medina, the vendors of the house
had no bank account. Our attempts to procure a bank draft which
could be cashed over the counter failed when the bank told us it
needed to be made out to a specific name. We rang Larbi, who said
he had no idea what the vendors were actually called.

'So what do we do?' I asked, wondering why he was getting
such a healthy commission when he didn't even know this most
basic of facts.

'You should get the money in cash,' he said.

After further delays, we were presented with a huge pile of
money. We had nothing to put it in, and the bank couldn't produce a single plastic bag, so the three of us stuffed our pockets
with wads of notes and waddled out looking as though we'd
done a heist.

Back at the house, the elderly vendors were waiting in the *massreiya* with the scribe and his assistant. The old man sat cross-legged
on the floor, dressed in a dark brown djellaba, the milky disks of
his sightless eyes reflecting the light from the window. The bearded
scribe was shouting alarmingly into his ear and gesticulating with
a roll of parchment that resembled one of the Dead Sea scrolls.
The old man's wife looked worried.

'What are they saying?' I fretted to Sandy. In the wee small hours I had gone over my doubts about the wisdom of the venture, but now I just wanted it to happen.

'No idea,' he shrugged, struggling to focus through his feverish haze.

Just when we needed him, Nabil had disappeared with Larbi to have a look at the house, so we could only guess, and wait.

Down in the tiled courtyard, two sheep waited their fate at *Eid al-Kbir*, the Feast of the Sacrifice, in three days' time. Callously I hoped that their blood wouldn't stain our tiles. In the orange tree just outside the window, small birds hopped about, chirping in the cool winter sunshine.

Nabil returned. The scribe was still shouting at the old man. 'Can you find out what the problem is?' I urged.

He listened for a moment and then shrugged. 'He's just telling him he needs to pay his back taxes before the sale can go through.'

That sounded harmless enough. I breathed a sigh of relief. At this point I would have paid the taxes for him.

Finally the scribe began to read a description of the property, which might have been a verse or two of *A Thousand and One Nights* for all our knowledge of Darija. Nabil didn't translate, but simply inclined his head. When the scribe had finished intoning there was a pause, and he turned his gaze expectantly to us.

'Please tell the owner,' I told Nabil firmly, 'that we're buying the house as is and nothing must be removed.' We'd met expats in the Medina who'd had windows and doors stripped by the owners after the completion of sale. These items fetched big money on the antique markets of Paris and London, and we knew that the ceiling in the *massreiya* alone was worth more than the entire house.

When my instruction was translated the old man looked pained. 'Tell her,' came the translation, 'that if you find something embellished, then it won't be unembellished.'

The issue of vanishing artefacts was a pervasive one. A friend of David's had recently bought a magnificent domed ceiling for his Fez house from an antique shop in Rabat. The workman installing it recognised it as one stolen from a house two doors down a few months before. The police were told, and the antique shop had to give the money back and pay for the reinstallation in the original house.

But many inhabitants of the Medina had little else to sell but their doors and windows, so why put artefacts before welfare? David's argument was that destruction of cultural heritage was short-sighted, and far greater numbers would be employed by tourism in the long run.

It struck me that for people living in a city more than a thousand years old, their lifespan represented a brief flash. It was for their descendants' sake that things needed to be preserved, but how do you deny anyone the right to modernise? Must they sacrifice their aspirations for the new because they live in an historically significant city?

After the signing, which for the old couple amounted to a thumb pressed into a red stamp pad and then onto the paper, Sandy, Nabil and I unburdened our pockets of the bundles of dirhams and handed them to the scribe. He counted the notes, then his offsider counted them again before handing them to the blind owner. As the pile of money in the old man's lap grew, so did his smile.

'It's the most money he will ever have in his life,' Nabil said. 'And it's the last time he will see it.' The couple were buying an apartment in the Ville Nouvelle and were looking forward to a bright, shiny new place where everything worked. No doubt they thought us deluded for wanting to buy their decrepit old house. I must confess that the same thought had crossed my own mind.

The mediaeval transaction finally done, a jug and glasses appeared. 'Great,' said Sandy, shivering with a temperature. 'A hot drink.'

Actually, no. We toasted our new house with cold almond milk and cookies.

'The owner wants to know where you come from,' Nabil said, and when I told him the old man repeated the name with wonder. It was as though a pair of Martians had dropped in to buy his house.

The old couple weren't due to move out for six months, and when the time came to pay the second half of the purchase price, it inevitably proved more complicated than we anticipated. Back in Australia, we opened our Moroccan cheque book, which we hadn't looked at since receiving it from the bank, and realised we had a problem. We stared at the incomprehensible Arabic, and the more we stared, the more confused we were about what went where.

Then it occurred to us that we knew one Moroccan who lived in Brisbane, but it transpired that he had left Morocco before he was old enough to have a cheque book, and he was equally in the dark. 'My brother back in Morocco is a notary,' our friend said. 'I could ask him.'

I scanned a blank cheque and emailed it to our friend to forward to his brother. Instructions were eventually conveyed to us and I made the cheque out to the vendors. I posted it to Nabil and sent another letter to the bank manager in Fez, along with a photo of the old couple so that he could identify them and pay them the cash.

As we had requested the vendors move out before the rest of the money was handed over, Larbi offered to organise a guardian to live in the house until I could return. At a cost of one thousand dirhams, the equivalent of about a hundred and fifty dollars, it seemed a good arrangement, especially since we didn't want to risk losing the fabulous doors and ceiling.

I planned to go to Fez in August to begin the restoration. It was impossible for Sandy to get time off work then, but we would be returning together seven months later. This wasn't ideal, but at least our dream house was finally becoming a reality.

'We can call it Riad Zany,' said Sandy. Zany was my childhood nickname, so I was touched. It wasn't particularly Moroccan, but it seemed appropriate for this eccentric venture. I was under no illusions that the restoration would be straightforward, but with patience and determination I felt we would manage, *inshallah*.

The final leg of my journey that August, from Dubai to Morocco, felt endless. It was around midday when I landed in Casablanca, and there being no plane to Fez until eleven that evening, I took the train. It rolled through the sprawling suburbs of Casablanca to the dry and dusty fields beyond. In outlying villages modern, whitewashed villas with satellite dishes rubbed shoulders with huts made of palm fronds, where plastic plugged the worst of the holes. Impoverished farmers with a donkey or two and a few chooks were trying to eke out a living on patches of parched soil. Freshly turned fields appeared to be sprouting crops of plastic bags, which had blown on the wind and been ploughed in.

Then the land began to form into soft rounded shapes as the foothills of the Atlas began. Vineyards and olive groves flicked past the window, then at last I could see the outskirts of my adopted city, nestled beneath a foggy haze of wood smoke.

I took a taxi to the Medina, gazing with pleasure at the mass of humanity flowing around me. The natives of Fez – Fassis – looked mediaeval in their djellabas and headscarves, although the colourful clothing of Moroccan women gave them a visibility on the streets that was in stark contrast to the crow-black invisibility of their counterparts in Dubai.

I collapsed into my hotel bed as soon as I'd checked in, and by seven next morning I was at the Café Firdous opposite with a *nus-nus* (half espresso, half milk) and a croissant. Café culture in Morocco is traditionally male-only, but in recent times the locals have become accustomed to strange foreign women who see no shame in sitting on the street sipping coffee. I watched the men greet one another with ardent kisses on both cheeks and cries of *'Salaam Aleikum!'* (Peace be with you!) To this came the response *'Aleikum Salaam!'* (With you be peace!)

Breakfast done, I strolled around the Medina, waiting for a civilised hour to collect the keys to the house. My heart lifted at the sight of a man on a donkey coming in under the Bab Bou Jeloud. Ironically, this magnificent Islamic keyhole-shaped gate, the most photographed monument in Fez, was built in 1913 by the French, to give their troops direct access to the two main streets of the Medina. The Bab Bou Jeloud replaced a copy of the Arc de Triomphe, which the French had previously built on this spot. Realising the locals would view this as provocation, a subsequent colonial administrator had it torn down and replaced with something more culturally appropriate. The gate is blue on one side, the colour that represents Fez, and green on the other, the colour of Islam.

After buying a new card for my Moroccan mobile, I tried to call Nabil, only to get a message saying his line was 'no longer open'. I rang his office and learnt he'd emigrated to Canada. David was unable to help, as he was in the United States. Braving it, I rang Larbi, who managed to understand my basic French and said he'd meet me at four that afternoon at the hotel. I couldn't understand why he didn't just have my keys dropped off, but my French wasn't good enough to argue.

In the meantime I decided to go to the house anyway and see if the guardian Larbi had hired was there. Pleased to find that my memory of the Medina was still intact, I made my way through the maze of streets to my new doorstep, knocked and waited. Then I knocked again, and kept knocking until my knuckles were sore. No one was there.

I was about to admit defeat when the door opposite opened and a plump young woman poked her head out. I introduced myself and explained I was looking for the guardian. She didn't speak French, she told me in halting French, and began talking in Darija, gesturing me to come inside. I smiled, nodded and followed her up a set of stairs into a tiny flat, part of a dar. She told me her name was Khadija, ushered me into a chair and went off to make tea, returning with her son Ayoub, an adorable five-year-old with enormous black eyes and a mischievous manner.

Another two women came in and greeted me, kissing me once on each cheek then three times on my right cheek. They were Khadija's sister and niece, and began telling me things in a mixture of Darija and French that I didn't fully understand. I explained the situation as best I could, and Khadija's brother-in-law was produced. He spoke much better French and insisted on ringing Larbi to see if he could come earlier. But the reply was still four o'clock.

My new friends invited me to stay, but not wishing to impose, I said I was going to take a walk in the Medina. Khadija wanted to come with me, but first she had to get ready. She spent a considerable amount of time changing, reappearing in a loose white suit with black pinstripes. Then she began, with great care, to apply her makeup. She made a fuss of doing her hair, and then covered it all with a scarf.

While she was getting ready I had a look around the room. Besides the banquettes – the long low Moroccan couches used for sitting on during the day and sleeping on at night – there was the ubiquitous television, and a display cabinet filled with photos, including one of Khadija at her wedding to a moustached fellow. Over the door was a photo of the Prophet's tomb at Mecca, surrounded by pilgrims, and in every crevice of the room were vases of fake flowers.

Eventually we left. Out in the Medina many shops were shut, it being Friday, the holy day. Most of the streets were too narrow for us to walk abreast so I hung behind, but as we passed a small photo studio, Khadija slipped her arm through mine and told me she'd like a photo of us together. I was both flattered and amused. We had only just met but already she seemed to regard me as her close friend. Briefly I wondered if there was more to the gesture than friendship; I smiled but made a noncommittal reply.

We made our way to a part of the Medina I had never seen

before, where I met Khadija's brother, who was selling bunches of luscious grapes. He had a handsome face but opened his mouth to reveal a double row of rotting teeth. Tooth decay is rife in Morocco because of the amount of sugar people consume. Every glass of mint tea has about six sugar cubes in it and the locals drink it incessantly.

Khadija's mother worked at another stall further down the street, and on being introduced promptly invited me to lunch.

We followed her to a tiny, three-room flat, stuffed with shabby furniture. The spaces between the furniture were taken up by women on their hands and knees washing the floor. I was ushered through into the lounge room and seated, then the women came through in succession for more greetings and kisses. For a few moments I felt like visiting royalty.

The women were curious to know everything about me, including how much we'd paid for the riad. I made the mistake of telling them, and thereafter it was conveyed to everyone who entered the room. Perhaps it gave them a sense that their own property was worth something if these crazy foreigners were prepared to buy the dumps they were desperate to get out of. It also seemed important that my husband was a radio journalist. That I was a journalist for a newspaper appeared to have little significance, and as far as I could make out was not one of the pieces of information passed on.

The incessant drone of the television played in the background, broadcasting pictures of men praying in a mosque, accompanied by religious songs. It reminded me of those Sunday-morning

Christian programs on Australian TV, but probably had far more of a following. It was also a way for Moroccan women to keep up with a service from which many were excluded. Only women past child-bearing age are permitted to go to mosques in Morocco, and then they worship in a separate area, behind the men. Young women must avoid any contact that might lead to sexual attraction, and therefore pray at home.

I helped Khadija's mother strip a bunch of mint for the tea. She asked how many children I had, and when I replied none she gazed at me with sorrowful eyes and patted my hand. Knowing she was unlikely to understand that it was a choice, I gave a look of resignation. 'Wow!' I said when she told me she had six. My exclamation was repeated, with general hilarity, to every subsequent guest who arrived.

Ten of us crowded into one room to eat, including Khadija's two sisters, her father, grandfather and a couple of sisters-in-law. I was offered eating utensils, but declined so as to fit in with everyone else. A kettle of water was taken from person to person and poured over outstretched hands, but the towel that followed was grey and greasy. Oh well, I thought, I need to build up some resistance to the bugs. On the table were rounds of freshly made wholemeal bread, which we tore and dipped into a communal platter of meat, chickpeas and gravy. This was accompanied by bowls of salad. To my discomfort, Khadija's mother waited on us while we ate, then did not join us.

I had a hard time identifying the meat. It looked like the knee and shin bone of — could it be a horse? I tried not to dwell on it, especially as Khadija insisted on putting morsels of the gelatinous stuff in front of me. It was very sweet of her, but I had to force myself to eat them. As soon as I ate one, more would appear, until I remembered to say, '*Safi*,' the Darija word for sufficient.

Khadija's mother wanted me to stay and rest with the others after lunch, as was the custom, on the banquettes in the adjoining room. But I needed to return to the hotel to get my keys, so I gave my apologies and thanks and left.

Larbi arrived on time, smartly dressed in a new white woollen pyjama suit with piping around the neck and arms, and new yellow slippers. Business was obviously booming.

'Yes,' he agreed when I asked him if he'd been busy. 'So many foreigners are buying houses.'

He led me on an unfamiliar route to our riad, through a labyrinth of streets so winding and complicated I was completely disoriented by the time we arrived and hardly recognised the place. We had come to the back entrance, an unprepossessing small metal door that led to the upstairs floor. It was on a completely different street, at right angles to the one on which our front door opened. The two streets were walled off from one another, and getting to the main entrance meant walking around the entire block.

As Larbi opened the door a flurry of dust arose. And not just at the door – it coated everything, particularly the floor in the room with the beautiful ceiling, on which I was further dismayed to see pieces of fallen plaster and debris. The grey feathers of some unfortunate bird lay scattered about and a few things were broken, including one of the steps in the entranceway. It didn't look like anyone had been there too recently, but at least the electricity still worked, though the wiring appeared dodgy and makeshift.

The house seemed smaller than I remembered, but lovely all the same, despite the dirt and disrepair. I was elated, and couldn't wait for it to be clean and liveable.

The following morning I woke at two a.m., excited to think that this was the day I moved in. Just after seven I was walking with a spring in my step down through the main artery of the Medina, the Tala'a Sghira, with only the street sweepers and the odd, early-delivery donkey about.

As I struggled with the unfamiliar keys of the riad, I noticed a few bags of builder's rubble beside the front door and wondered what they were doing there. Inside at last, I wandered round checking things out. I turned on the water in the kitchen, something I hadn't done the previous day. It flowed for a few seconds, slowed to a trickle, gurgled and stopped. It was obviously off at the mains, and I realised I probably had to pay a bond or something to the municipal authority.

Next I took the stopper off the squat toilet and got a shock. In the hole were brown granules, which were not of human making. I hoped they were simply the result of the toilet not being used for a while. Covering my hand with an old plastic bag, I scooped a few grains out. It was dirt. The hole was full of dirt, and appeared to be unconnected to the sewer. Suddenly the bags of rubble next to the front door made sense. Perhaps in a last-ditch attempt to fix the toilet, the previous owners had dug out the sewer connection.

Obviously no one had used the toilet since the old couple had left. So much for Larbi's guardian. Someone had been living here for two months who didn't piss? Let alone anything else. The toilet was the one thing Nabil couldn't have been expected to check before he paid the remainder of the money. My spirit of optimism turned a little sour.

When Larbi turned up a short time later with a couple of cleaners, I pointed out that there was no water and he sent them away. I showed him the toilet and asked how this could be, if a guardian had been living here. He was adamant that someone had been. 'He worked during the day,' Larbi told me.

Since there'd been a trickle of water when I tried the tap, it seemed unlikely that anyone else had tried it after the water was turned off. Nevertheless Larbi pressed me for the money to pay the guardian. I deflected, saying we would talk about it once the toilet was fixed.

I went with Larbi to the electricity and water company in the nearby R'Cif district and paid the bills for the previous owners. The amount wasn't huge and I was too desperate to have the water back on to worry about the fact that they weren't actually my bills. But then I learned that I needed proof I owned the riad before the utilities could be reconnected in my name. The scribe who had done the transaction for the house would have to give me a document verifying ownership.

Outside on the pavement, Larbi rang the scribe, only to find that he was on holiday hundreds of kilometres away and wouldn't be returning to Fez for a week. I envisaged an expensive hotel bill; not only that, but two Australian friends were due to arrive shortly and expected to stay in my house. I had blithely invited them several months before, not really thinking they'd show up, but now they'd be here in a few days. If they weren't able to stay with me, it would only be polite to pay for their hotel.

The office was about to close and there was nothing that could be done immediately, so Larbi and I arranged to meet again on Monday morning and try to find a solution.

I made my way despondently back to the riad, where I couldn't even make myself a cup of tea to mull the situation over. As there were no chairs, I was perched on the edge of the fountain, pondering what I could do, when a sharp rap sounded on the door. Opening it, I found a dapper man with an official-looking navy cap and a French leather satchel slung over his chest. He looked as if he'd just stepped out of a Jacques Tati film.

He was the utilities collector, he explained, and presented me with two bills made out to the previous owners — the very bills I had just paid. Such personal service was unexpected. In Australia you were simply sent a notice, then a reminder, and if that didn't work you got cut off. There was no opportunity to stand around discussing it.

I managed, after three attempts, to convince the man that I'd paid the bills only that morning, then went and knocked on Khadija's door for a shopping trip we'd arranged. I relayed to her as best I could the situation with the toilet and she told me her husband Abdul could fix it once the water was connected. He could also do the painting and plastering, while she was more than happy to do the cleaning. I tried to explain that I was hiring a plumber, but somehow that got lost in translation – or in her desire for her husband to have some work. I didn't mind Khadija doing some cleaning, but I didn't want to employ her as a regular cleaner. Living in such close proximity, it would be too difficult if things went wrong.

Khadija and Abdul and little Ayoub were all wearing their best clothes and I realised with a sinking heart that they too would be coming shopping. I had been grateful when Khadija offered to take me, as I had no idea where to go, though it did cross my mind that she might be planning on receiving a commission. Three of them was a far more unwieldy proposition, but how could I say no? So off we went.

We walked down to the road at R'Cif, one of the few places where cars could access the Medina. Catching a taxi wasn't easy; the turning circle was crowded with people who swooped on any arriving vehicle, fighting to get in before the previous occupants had got out. There's no such thing as a queue in Morocco, and the quickest person claims the taxi. When we finally got one Ayoub had to keep his head down the whole way as it's illegal for a petit taxi to carry more than three passengers. Petit taxis are one of the major forms of transport around Moroccan cities, and in every city the colour differs. In Fez they are red, in Marrakesh beige, in Rabat blue.

I had a list of things I needed: a toilet, a bed, stove, kitchen things. Khadija was amazed when I said I didn't want a television, an item that was obviously an integral part of her family's life. Abdul directed the driver to an enormous supermarket on the outskirts of the Ville Nouvelle, a palace dedicated to Western luxury. The size of a football field, it was packed with household appliances, furniture, clothing and food. But no toilets. I bought everybody an icecream and they wandered around the aisles wide-eyed. It occurred to me that they had never been there before.

It became clear that Abdul didn't have a clue where to buy a toilet, which didn't give me a lot of faith in his expertise in installing them. I rang an expat I knew who redirected me to another district in the Ville Nouvelle, which had a street of plumbing-supply shops. There I found a suitable loo, whereupon Abdul tried to impress me with his bargaining prowess, rejecting the owner's insistence that it was a fixed-price shop and arguing until he had us thrown out.

We walked along to another shop and I spotted a French toilet I liked. This time I took over the negotiations and bought it. I had less success organising delivery, though, and instead arranged for Abdul to pick it up on Monday. He seemed to vacillate between being dopey and aggressive, and I realised I'd need to limit his involvement.

The shopping done, I gave my neighbours the money for a taxi and split from them with relief. I had imagined a nice girly time with Khadija, and had planned on taking her to lunch, but the entire family was hard work.

The following day was Sunday and Khadija had asked me to

go to the *hammam*, or bathhouse, with her. I went to collect her and found her mother, aunt and sister there. Khadija brought mint tea and the conversation rose and fell in Darija around me, only the odd word making sense. I peered through the window into the courtyard below. Khadija had told me that four families lived in this dar, which was smaller than our riad but had many more rooms, all of them tiny.

The women asked if they could see my house, so we trooped across the alley and they went from room to room, cooing, '*Une belle maison.*' It must have been unimaginable to them that just two people were going to live here.

Khadija was pleased to discover that I had purchased some cleaning equipment, but when I asked how much she charged for cleaning she said it was up to me to decide. I had no idea. I knew that a master craftsman was paid a hundred and fifty dirhams per day — about twenty-two Australian dollars — but Khadija received a pittance for her work embroidering sequins on slippers, so whatever I paid was going to seem like money for jam.

I wanted to be generous, but I had both read and been told that if you paid much more than the going rate in Morocco it was taken as a sign of gross stupidity. I didn't want to set up unrealistic expectations either, so perhaps I could give Khadija a gift in addition to a payment slightly better than usual.

Khadija, little Ayoub and I walked down to the *hammam* in R'Cif. This was an old, poor area and the bathhouse had no pretensions. It had been in continuous use for around four hundred years and its walls, although whitewashed, were stained with patches of green mould. The tile work on the floor was so worn that most of the colour had rubbed off, and there were layers of soap scum around the edges. There was an anteroom for undressing, then two large adjoining rooms, each with a barrel-vaulted ceiling punctuated with round holes through which shafts of light poured, spot-lighting semi-naked figures in a haze of steam.

The first of the rooms, the cooler of the two, was the domain of mothers and small children, while the inner sanctum was for young, unmarried and older women. We chose a spot in the first room, in a small alcove upstream from everyone else, and Khadija left Ayoub with me while she went to fill the plastic buckets.

The contrast between the traditional modesty of Moroccan women in public and the relaxed intimacy of the *hammam* was star-tling. I could see cellulite, flabby breasts, bodies of all shapes, sizes and skin tones. (Many Moroccan women seemed to take particu-lar care with their eyebrows, plucking them to a beautiful arch, perhaps because, with limbs and heads covered up on the street, eyebrows became a more distinguishing feature.) They kept their underpants on, but chatted as they lathered themselves, each other and their children. There were a few cries when shampoo got into small eyes, but mostly the children seemed to relish the attention. One little girl sat in a red bucket bigger than she was, her legs dangling over the edge.

Through the archway, I glimpsed Khadija mixing water from the fountains of hot and cold gushing out of the wall. Satisfied she had the temperature right, she slid the buckets across the floor, settling down next to us. We proceeded to soap up with a soft, translucent brown substance that resembled axle grease but which was soap made from olive oil. Taking a rough black mitt, Khadija scrubbed herself, then a squirming Ayoub, and finally me. It felt like being licked by a mother cat.

We had dressed and were about to leave when an argument broke out between two women in the first room. It quickly escalated into a shouting match, with others joining in. The two protagonists had their faces close together, hands shoving one another's chest. The shouting, jostling and anger raged for several minutes. I longed to know what it was about. Jealousy over a man? A friendship betrayed? An agreement broken? But Khadija just shrugged her shoulders. Perhaps, I thought, when there is no concept of private space, as was the case in Morocco, in public is as good a place as any to express frustrations.

Walking back with a towel over my wet hair and Ayoub racing happily ahead, trying to join in the big boys' game of street soccer, I felt relaxed and at peace. Communal bathing is a lovely tradition. I was almost going to tell Khadija about the saunas Sandy and I had with friends every Sunday night, but thought she might be shocked that they included both men and women. She and I had managed to develop a reasonable communication, with me speaking in simple French and Khadija replying in a mixture of French and Darija. If the subject was not too complex we seemed to understand one another, but we were far from being able to have an abstract discussion about the very different lives of women in our respective countries.

I woke on Monday morning and lay staring at the walls of my hotel room, mulling over the problem of proving my ownership of the house. The thought of being stuck in the hotel until the scribe returned from holidays was not a happy one, but I couldn't see an alternative.

At nine o'clock I went to meet Larbi at the utilities office. There was a crush of customers fighting to pay their accounts – no standing in line, just dozens of people trying to attract attention by waving their bills and shouting at bureaucrats who went about their business with bored expressions, studiously ignoring the mêlée. This could take hours, I thought. We were at the back of the crowd and I knew I was never going to get used to the Moroccan method of elbowing one's way through with brute force.

Larbi, however, had no such qualms and joined in the shouting, calling to someone he knew behind the counter. A man sauntered over and Larbi explained my predicament. Then, translating to me, he confirmed what I'd been told on Friday – it wasn't possible to officially reconnect the water without the paperwork.

Bugger.

'But he has also told me how to turn the water on in the meantime,' Larbi continued. 'You just need a plumber.'

Hallelujah.

Back in the Medina we found a plumber's shop. Larbi had a few words with the elderly, bow-legged proprietor, and after collecting a spanner the man waddled with us to the riad, where we went around to the back door. To one side was a small metal flap I hadn't noticed before. The plumber lifted this, gave a few twists of his spanner, and, miracle of miracles, the sound of running taps could be heard through the house.

'Yahoo,' I yelled loudly, to the amusement of Larbi and the plumber. I raced from room to room, turning off the water that was spilling all over the tiled floors.

I paid the plumber generously for his ten well-spent minutes and he left a happy man. Realising that now was a good moment, Larbi asked again about payment for the ghost guardian. I was feeling so buoyed that I almost paid him without fuss, but something held me back. If there had been a guardian why hadn't he told Larbi there was no water in the house? How could he have lived there for two months without any water or a working toilet? It was beyond belief that there'd be no give-away smell of pee in the courtyard, as I knew it hadn't rained at all during that period.

Then there were the letters I'd found piled against the front door, as though it hadn't been opened in a long time. There was no sign of footprints or a mattress in the thick dust that coated the floors. And there was one final, irrefutable piece of evidence. Khadija, I had discovered, was an extremely curious neighbour, yet she had never seen this guardian. Not once.

By way of reply I requested the guardian's telephone number. Larbi declined, saying I should pay him instead. I said there were some questions I wanted answered and I needed a translator as my French was so poor. It would have to wait until David returned.

Larbi departed and I knocked on Khadija's door to tell her I had water. Leaving her in the riad to begin cleaning up the dust, Abdul and I went back to the Ville Nouvelle for the toilet. I also bought a small hand-basin, a tap, and a hose for washing one's bits. A small delivery van drove the goods and Abdul to R'Cif, but there wasn't enough room for me so I followed in a taxi. When I arrived Abdul had already organised a donkey to transport the stuff to the house. The sight of my fancy French loo riding on top of the donkey was surreal. The only thing weirder would have been someone sitting on it.

Khadija had done an amazing job of sweeping and mopping the entire downstairs, and Riad Zany was finally beginning to look cared for. The toilet unloaded, Abdul and I returned to the plumber's shop, then accompanied him to the hardware store to buy the fittings he needed to install the unit. The hardware shop was a hole in the wall, piled to the ceiling with everything from locks and screws to rope and wire, bags of cement and tools. The plumber reeled off his requirements and the shop assistant ran up and down a ladder fetching them. It was a world away from the hardware megamarkets that proliferated at home, where you could spend an hour traipsing up and down aisles searching for what you needed, and even more time finding someone to help you.

To my great relief, the plumber began work the following day. His first concern was that he might damage the beautiful *zellij* in the tiny bathroom. I was reassured that he cared about this, but my need for the toilet was greater than my aesthetic sensibilities at this point and so I shrugged and said I would get it redone. By the end of the day I had a functioning hand basin and tap. Things seemed to be progressing well and I felt confident the work would be completed next day.

But as usual, I hadn't allowed for the unexpected. On Tuesday, Abdul, who was lending a hand to dig out the old toilet, gave a cry of exasperation. He had dug down to the point where they hoped to connect the new pipe to the old, and there was no old pipe at all.

I didn't understand why this was such a problem. Why couldn't they just keep digging and reconnect with the sewer? At the limits of his second language, Abdul went to fetch the owner of a local restaurant, who spoke better French.

The restaurateur explained that the sewerage in Fez was the oldest functioning system in the world. When our house had been built centuries earlier a trench system lined with tiles was used, instead of pipes. Over the years the trench had narrowed and collapsed in places, so that now only a trickle of water could pass through to the main sewer line in the street outside. The trench was not nearly wide enough to allow the rush of water generated by a modern toilet.

It seemed the only solution was to widen the trench. Unfortunately, this ran out to the sewer line right under my front stairs, and digging them up would mean losing all the complex tile work.

Disappointed at the thought of several more days in the hotel, I left them to it and went and had a cup of mint tea with Khadija. It was the first time I had been in her kitchen, which was incredibly dark and dingy and was shared with three other families. She told me they'd been living in the house for five years and paid six hundred dirhams a month in rent, about ninety dollars. As they earned so little, it must have been a bit of a stretch.

Then she hit me with it. Trembling with nervousness, she asked if she, Abdul and Ayoub could move into my house when I went back to Australia.

From her point of view it must have seemed perfectly logical. Here was this big empty house across the alley, and while she was cleaning she must have daydreamed about living there. In some ways it was a good solution to let them stay, as I needed to find a house minder – the problem would lie in getting them to move again when Sandy and I returned. I'd been warned numerous times that if people decided they wanted to stay, it was very difficult under Moroccan law to get them out.

Suddenly I understood why her relatives had been trooping through the riad in the past couple of days, checking out the place. While the cleaning and bathroom work had been going on, numerous aunts, sisters and their children had appeared, wanting to look around. I suspected that my plane would hardly have left the tarmac before there'd be about nine of them living there.

I felt awful shattering her dream. I said gently that it was very nice of her to offer but I had made other arrangements. She kissed me and said she understood, but I could see she was disappointed.

For the rest of the day I carried an uncomfortable sense of guilt. I was so much wealthier than my neighbours, and Khadija had started calling me Madame Suzanna, which made me feel strange. From her perspective, it defined the social difference between us, and although I protested she persisted.

Something else I found disturbing was dealing with beggars. There is little social welfare in Morocco, so the unemployed, the disabled, divorced or widowed women, the latter usually with children and no job skills, are often left without an income. While I frequently gave money to beggars, it wasn't feasible to give to everyone who asked, and selecting who was most deserving could be difficult. Once, having just given to an old lady with a goitre problem and a man with no arms or legs, I was accosted outside my riad door by a man holding a sleeping child. He said he wanted money to buy milk for her. As I was about to give him some, he told me he needed seven euros, claiming that was the cost of the milk.

Really? A couple of dirhams was more like it. A vision of him appearing at my door every time I opened it flashed into my head, so I told him no, then felt tremendously guilty for hours afterwards. I knew that Moroccans deal with situations where they choose not to give by saying, 'May Allah make it easy on you.' It was a phrase I needed to learn how to say.

Deciding to take a positive approach and check out of the hotel, I packed up my belongings and moved to the riad, where my optimism was rewarded. Abdul had managed to widen the trench sufficiently without digging up the stairs after all. The plumber spent the morning fiddling with the insides of the toilet and the pipes, and around eleven o'clock I heard the sweetest sound in the world – my toilet flushing for the first time. Before that moment, if you'd told me that such a noise would produce a rush of pure ecstasy, I'd have said you were bonkers.

A week or so later, I passed by the hardware store and saw my old squat toilet for sale. I knew when I set it outside the front door that it would find a new home; something so useful would never be wasted in Morocco. Although I considered myself frugal, it reinforced how much more I squandered than my neighbours.

Khadija arrived to help me unpack some new deliveries, including mattresses, and when everything was organised I paid her. This time, as she'd only done a couple of hours' work, I gave her fifty dirhams. This was as much as Abdul got for an entire day as a parking attendant, yet I saw disappointment on her face, and realised I'd got the amount wrong with the initial payment I'd given her for the big clean.

Once she'd left, I stood in the courtyard drinking it in. At last I was living in our house. I felt a rush of happiness bubble up.

It was late afternoon, birds were singing in the citrus trees, and the light was golden on the wall. I went into the downstairs salon and gazed out at the fountain framed by the plasterwork arch, then wandered upstairs.

Off the catwalk that joined the two main sections of the house, a set of stairs led to a tiny room whose purpose was a mystery. The ceiling above the passage was painted in geometric designs and on the end wall was Arabic script. When five-year-old Ayoub had visited he proudly announced that this read *Allah Akbar* (God is Great) and *Bismillah* (Praise be to God). Perhaps the room had been a place to pray in a busy household.

In the *massreiya* I threw back the shutters and light flooded in. The radial design on the ceiling had been painted so long ago that the colours had faded to beautiful subtle tones. At that moment, not even the amount of work involved to fix the sagging on one side spoiled my appreciation of it.

I looked at the date on the intricate band of plasterwork bordering the ceiling: it read 1292, the Muslim calendar year for the date of the last major restoration to the house. That converted to about 1875 in the Western calendar, and if the house had stood for that long without further maintenance, surely it would for a while longer. At least until we could fix it.

That evening, I went out to eat in the Medina, returning after dark. It felt eerie being in the riad by myself. Before going to bed, I investigated all the rooms, shining a torch into every dark corner to reassure myself I wasn't going to get any unpleasant surprises during the night. I had made the downstairs salon my bedroom, and as I lay down I found it comforting to hear muffled voices through the walls. I fell asleep to the buzz of mosquitoes and a persistent blowfly.

I slept lightly, waking at the fall of a leaf from the lemon tree, and again when tomcats had a territorial stoush on the terrace. In the early morning I was roused by the sound of a mournful song, and a short time later came a muezzin's call. This went on for half an hour, joined by competing chants from other mosques. Just as they finished a cool breeze drifted in, displacing the hot and heavy night air, and I slept well for a couple of hours.

The next time I woke it was to the sound of my mobile phone. It was Sandy, who'd tried to call on three other occasions and got a Moroccan who spoke no English. Odd, given he'd dialled the same number each time. Sandy had just spent several weeks in Sydney, filling in for another radio presenter who'd quit unexpectedly. He was back at home now, with the cats curled up on his knee. I felt a rush of love for my small family. Some men might have felt resentful and paranoid at having their partner out of sight for so long. The truth was, I hadn't met anyone in years who came close to touching my mind and my heart the way Sandy did. Nor was I looking to.

Buoyed by my success with the small bathroom, I decided to get started on the fountain. It was quite a few years since it had done anything but collect leaves, and the roots of the trees on either side appeared to have interfered with the water pipes.

I fetched the old plumber and immediately ran into communication difficulties again. As with the toilet, the problem with the
fountain was greater than it looked, and understanding it required a
translator. The restaurateur came to the rescue once more, explain-
ing that the plumber was at the limit of his skills, and wasn't about
to start removing the beautiful *zellij* around the fountain.

Zellij is like a jigsaw puzzle; the pieces are built up progressively
into a pattern. You can't simply take some out then easily repair it,
and at the equivalent of thirty dollars a metre it was, in Moroccan
terms, extremely expensive. The restaurateur smiled and said he
would send me a real plumber, who was a good Muslim. I guessed
that meant he would be respectful and wouldn't try to cheat me.

Unable to do anything further, I strolled to the souk and
bought some luscious peaches for Khadija and her family, which
cost about half Abdul's daily wage. It was Friday and I'd been
invited for lunch.

While we ate — chicken with noodles, a tomato and onion
salad, fresh fruit to follow — I asked how she and Abdul had met.
He'd first glimpsed her when she drove in to park at his parking
station, he told me, his unshaven face lighting up as he spoke.
Somehow he'd mustered the courage to approach her, though I
gathered Khadija hadn't been too impressed with him at first. Perhaps she harboured higher aspirations — I didn't ask. But he won
her over and went to her family to ask for her hand in marriage.

When I enquired how she'd felt about this she glanced down
coyly. Her dowry of two thousand dirhams had been arranged,
and they had an elaborate wedding that went on for three days

and nights. Before it began, Khadija's friends had prepared her carefully, painting designs on her hands and feet with henna and dressing her in an elaborate costume, which was changed several times over the course of the ceremony. The bride and groom initially celebrated separately, then Abdul rode to Khadija's place on a white horse, accompanied by friends who sang and banged drums all the way. After more partying, bride and groom were seated on circular platforms covered with gold fabric, which were suspended aloft and carried into their bedchamber.

They produced a series of photo albums. Khadija was only seventeen at the time and the prospect of marriage must have been confronting, for in the pictures she looked grim, unsmiling and scared. Perhaps she was contemplating the moment when the wedding sheet with the spot of blood would be carried outside to the waiting crowd, a practice still current among all but Westernised Moroccans. If the bride does not bleed it shows she wasn't a virgin, and huge shame is brought upon her family. Her new husband may even disown her.

There were also photos of young Ayoub on his 'cutting day'. Circumcision is obligatory for Muslim males, and a major event in the lives of young Moroccan boys. It is done between the ages of three and seven, when the boy is old enough to remember the occasion but too young to make trouble. There are variations on this amazing ritual all over Morocco, but common elements are shared by traditional communities.

Before the chosen day, which is usually during spring, the house will be thoroughly cleaned and a room whitewashed for the boy's use. Special bread is baked and a ram purchased for slaughter. The slaughtering is done the day before, and blood is put above the front door to ask the blessing of the djinns, or spirits. That evening, the boy's mother places a wooden plate of sheep or goat excrement on the terrace under the stars, to endow it with magical properties.

A special bandage is sometimes prepared, by marking it with five vertical lines using the liquid from the saffron plant. This symbolises the hand of Fatima, who protects against the evil eye. Two bags containing a mixture of nigella seeds, salt, a tiny silver coin, and various other items are wrapped in red rags and attached to pieces of red wool. One bag will be worn by the boy, while the other one is hung somewhere in the house.

Early in the morning on the day of the circumcision, the mother will dress her son in a white djellaba and a green felt hat. A male relative, often the boy's uncle, will collect him on a white horse. The boy sits in front as they ride around town, followed by male family members and friends chanting religious songs. Holy men bless the boy before he returns home.

The circumcision is done by the local barber, who will be waiting at the house. The imam of the local mosque chants verses from the Koran, along with some of his students, then the women of the house and the female guests sing ritual songs to comfort the boy. In the meantime he is carried to the specially whitewashed room and the barber relaxes him by having a friendly chat. At the chosen moment, the boy is asked to lie on his back and put his feet over his head. A piece of sheep excrement is taken from the bowl and the little penis pushed through it, so that the tip is visible while being held firmly in place.

Then a favoured distraction technique is for the barber to point to a corner of the ceiling and tell the boy there is a cute little bird there, before whipping out his knife and cutting off the foreskin. The penis is cleaned with antiseptic and bandaged, then a rooster is brought to have its comb cut, the belief being that if only a single cutting is performed, it may bring evil to the boy.

While this is happening the women continue to sing and clap, and when the boy's cries are heard they increase their volume. He is carried three times around the circle of singing women, who ululate wildly.

Afterwards the boy is taken back to the whitewashed room and a huge fuss is made of him. He is served boiled eggs, lamb kebabs and sweets, and given gifts, usually money. The women return to the courtyard, still singing and clapping, and one of them will grab the wooden plate and throw the sheep excrement over the others, to bring fertility to any crops they have, along with general good fortune. The rest of the afternoon belongs to them. They may dress up in even more elaborate clothing and dance to music from an all-female group. Special food is served – a kind of donut, with dishes of butter and honey, followed by a tagine.

Later, the mother will carry the foreskin to the mosque in a bowl of henna, making a wish for her son to be devout and successful in society, before burying it.

In the first of the photos, Ayoub looked bemused but proud, as if he knew he was special on this particular day but had no idea why. In later photos he was downcast and miserable, and when Khadija passed these to him he burst out crying, rolling around with tears streaming down his face. His circumcision was obviously a painful experience he would rather forget, yet one his parents were proud to remember.

Having exhausted the photos, Khadija produced the family videos, but I pleaded tiredness, saying I hadn't slept very well on my first night in the unfamiliar house.

That was a mistake. Immediately Khadija said she and Ayoub would come and sleep in the riad to keep me company. Abdul joined in, saying it was fine by him. 'Ayoub would love it,' entreated Khadija.

I thanked them profusely and declined, thinking the matter settled, but when Khadija's sister arrived home a few minutes later it started all over again. The sister would also be more than happy to come and stay at my house, it seemed. And no doubt the sister's teenage daughters as well, I thought. My peaceful little house would be filled with women, and the men of the family would be running in and out visiting them. I tried not to look horrified.

Taking a deep breath, I explained as clearly as I could that I was a writer and needed solitude. I often woke up at odd hours in the night and wrote, I told them. It was my work, my livelihood, and it was necessary. Abdul understood and explained it to the others, and much to my relief the pleading ceased.

With no way of cooking in my basic little kitchen, I headed out for dinner again that evening. In the souk, crowds of people were going for their nightly stroll. The sweet sellers were doing a roaring trade in sticky mounds of deep-fried confectionery dipped in sugar syrup. I stopped to buy an almond-milk dessert, a bit like blancmange, served in a glass. It was cool, slippery and sweet on the tongue.

When I returned to the riad Ayoub was waiting at the top of the alley and called out excitedly. Abdul and Khadija were sitting on their steps with, wonderfully, a bottle of gas for me. They had even gone to the trouble of buying the hose and valve to fit it. I repaid them, touched that they had gone to the trouble. Now I could cook. As I didn't plan on having hot water in the kitchen, I'd bought an enormous copper kettle to boil for the washing up. But right now it was me who was getting the wash.

The kettle boiled quickly on the strong gas flame. I diluted it with cold water, poured it into a bucket, and carried it to the larger bathroom near the downstairs salon. Once, this long narrow room would have been a reasonable size, but now the bulging of the inner wall – the *flambement*, according to the engineer – had reduced its width considerably. The blue and white tiles were still appealing, but the shower and pipes were in an advanced state of decrepitude – rusty and useless. Like the work of a drunken plumber, they meandered all over the room in a most peculiar fashion, before heading ceilingwards to supply another tap upstairs.

I'd tested the shower previously, only to discover that the

water squirted straight upwards. It needed more than a new shower head; the pipes had to be completely redone, but I didn't want to do that before the *flambement* was repaired. Until then, I would have to make do with bucket washes.

But for every thoughtful act by Khadija and Abdul, there seemed to be one that left me feeling doubtful. One evening, they appeared at my door with a man they introduced as the guardian for the district. I had noticed him dabbing brown paint on door-ways lower down the alley in the past few days, but now Abdul said the guardian was offering to remove the bags of rubble left by the door by the previous owners. I was happy to have this done, and after some consultation with Abdul, a price of a hundred dirhams was quoted. That sounded pretty steep; the restaurateur had told me it would be around forty dirhams.

Khadija took me to one side and explained that, as this man was the guardian of the area, he was my second line of defence after them, so I needed to establish a good relationship with him. He also cleaned the street, she said, and if I paid him an additional twenty dirhams he would keep a special eye on my house.

I felt as if I were being ambushed. Was this some kind of strategy to extract money from me? What would happen if I didn't pay the money? Would the guardian let one of the neighbourhood thieves know when I was out? Feeling pressured I agreed, but only to paying half the amount now, and the remainder when the rubble was removed.

The street guardian started to finger my door thoughtfully, running his hands over the protruding metal studs. I knew he was just dying to be let loose on it with that tin of brown paint I had seen earlier. '*Non,*' I said, shaking my head. '*Non, merci.*'

Next morning the rubble was gone, and that evening, quite late, there was a knock on the door and three men I'd never seen before were waiting outside with Abdul. They wanted the rest of the money. I had no idea how they fitted into the picture, but I gave them the remaining fifty dirhams and they wrote me a receipt for twenty. Since I'd now paid a hundred dirhams, I told them I wanted a receipt for that amount.

They looked a little surprised, and Abdul quickly leapt in with an explanation I couldn't understand. They altered the amount and I closed the door knowing I'd been fleeced, but uncertain how they had divided the proceeds.

David finally returned to Fez, having spent August in the United States. I was on my way to catch up with him over dinner one evening when something unpleasant happened. Two young American women stopped me, saying they were lost and asking if they could follow me.

'Sure,' I replied. 'I'm just going down to where the taxis are.'

A moment later, a local man in his early twenties appeared. 'I will show you the way,' he said in French. It was a statement, not an offer.

'Thanks, but it's okay,' I replied. 'I'm taking them.'

He stared at me coldly. 'No,' he insisted, 'I will show them. The Medina can be a dangerous place.' He narrowed his eyes, clearly annoyed that I was getting between him and a potential fee.

'*Non, merci,*' I said firmly, not about to let him intimidate me.

He began to walk right in front of us, forcing us to slow our pace to his. I stopped to let him get ahead and the girls halted behind me, but he paused as well. 'I am showing you the way,' he said.

I told him that I lived in the Medina, I knew the way, and we did not need his help.

'I know exactly where you live,' he replied, his piercing green eyes staring right into me.

A chill ran down the back of my neck. I tried to keep my face neutral and move past him. At that moment he turned, stepping in front of me, and I accidentally kicked his ankle. Swinging round, he glared at me, eyes now filled with hate.

'*Excuse moi, monsieur*,' I said with a hint of sarcasm.

'You will follow me,' he stated, starting to lead us again.

I had had enough. '*Imchi*,' I said, an Arabic word which I thought meant 'keep walking'.

His reaction was explosive. A stream of Darija spat from his mouth. I didn't know their meaning but I had a feeling they detailed sexual acts involving donkeys and the origins of my birth. At least what I'd said had the intended effect, and he disappeared up a side alley. The girls were grateful but I remained disturbed.

I told David about the incident over dinner. We were at a formal, French-style restaurant in the Ville Nouvelle, which was empty when we arrived at seven o'clock, an early hour to eat in Fez.

'What does *imchi* mean exactly?' I asked.

'It's the Egyptian Arabic equivalent of "fuck you",' David said with a smile.

No wonder the guy was angry. 'Great. So I'm doing well

making friends in the neighbourhood,' I said. 'That's healthy.' I had a vision of being knifed in some dark alley on my way home, or coming back to find my laptop and camera gear gone.

'Look, the Medina is generally pretty safe,' David said. 'But something did happen last year.' He proceeded to tell me about a French couple who'd been walking past a mosque with their son and had exchanged a kiss right outside. Enraged at such Western sacrilege, some nutter ran out with a knife and tried to kill the woman. When the son defended her he was stabbed to death instead. 'The whole thing was hushed up,' said David. 'They didn't want it to impact on tourism. Anyway, the guy who did it was deranged.'

I knew he was right. Murders and assaults were rare in Fez; most crime was property-related, a result of poverty. David knew of a British couple, new to the Medina, whose house had been robbed. A neighbour had seen the burglars, and although reluctant to go to the police, he had given the names of the offenders to the couple. The police had already caught one of them, a fifteen-year-old who'd immediately confessed.

This was what came of living in a close community, I thought, where everyone knows everyone else's business. My house in Australia had been robbed a couple of times, and when the police eventually turned up they as good as admitted they were going through the motions, and there was little prospect of catching anyone.

The British couple were concerned about what would happen to the boy who'd been caught, and had considered not pressing charges. David assured them that Morocco had French and not

Sharia law, so the boy wouldn't have his hand cut off. (While some Moroccan fundamentalist groups advocate the introduction of Sharia law, the French system is so well entrenched that this is unlikely.) David thought it was better to let the police do what they had to, or word would get around that there were no consequences for stealing from foreigners.

Our food arrived – gurnard, a delicious fish – and the conversation moved on to Larbi and the ghost guardian. David looked speculative.

'Do you know something I don't?' I asked.

'Last year a couple paid him a deposit for a house. In the end the sale didn't go through, and of course they wanted their money back. Eventually Larbi came up with half of it, but he still hasn't returned the rest. He claims he gave it to the owners but he doesn't have any proof. He keeps saying he'll pay, then puts them off.'

So Larbi had money troubles. A small light shone on my own situation. It seemed that he'd pretended to employ a guardian in order to keep the payment for himself.

I told David my thoughts and he shrugged. 'I think Larbi is trustworthy up to a point, but I'd never tempt him. A few years ago I gave him the keys to my house to take care of it while I was away. When I came back I found a bottle of women's perfume next to my bed.'

'So he used it for a tryst?'

'There's nowhere here for couples to go,' David said. 'So they make use of wherever they can. I changed my locks after I got my keys back. Have you done that?'

I hadn't. I'd arranged for them to be changed after the final payment to the vendors, but Larbi had had my keys since then.

'If your keys have been in Larbi's possession, it's safe to assume he has copies,' David said. 'It'd be a good idea to change them before you have any kind of dispute with him.'

It was good advice. The last thing I wanted was to be involved in a conflict, but unless I was willing to pay up and shut up, it seemed inevitable. It was possible Larbi would simply come and remove what he thought he was owed.

'I think you need to insist on a meeting with the guardian and the translator, without Larbi,' David said. 'Don't pay unless you meet him and are satisfied he was actually there.'

This was along the same lines I'd been thinking, but it was reassuring to have it confirmed. I valued David's good sense and support.

I had never met anyone so wonderfully obsessive about old things as he was. He had started collecting antiques at the age of seven, going to auctions, fairs and antique shops with his mother. When I met his sister some months later she told me he'd been an odd child, always wanting to hang out with the old people across the road and hear their stories of days gone by, instead of playing with the other kids. I found this endearing. Even David's mobile phone had an old-fashioned ring-tone.

Next day, I called Larbi and said I wanted to meet the guardian. He didn't miss a beat, although not surprisingly he wasn't willing for the meeting to go ahead without him. I turned up at the appointed time, waited half an hour, but no one showed.

A few nights later, I was in bed reading when there was a knock on the back door. At first I ignored it. It was around ten p.m. and I wasn't expecting anyone at that hour. I wasn't about to open the door at night to someone I didn't know. But the knocking continued, louder and more insistent.

'Who is it?' I called out in French.

There was no answer, just more knocking, so I called out again. This time a male voice said something about wanting to come in and look at my house. What the hell for, at this time of night? I didn't think so.

'No,' I shouted. 'I don't know you.'

The knocking stopped and I breathed a sigh of relief. But a couple of minutes later it began on the front door. Since getting there required going round several alleys, the man obviously knew the layout of the house. This time I didn't answer. Then I heard Khadija calling.

'Madame?' Khadija and Abdul were having a conversation with whoever was outside.

'*Non*,' I replied. 'I don't know this man.'

Some time later I thought I heard someone climbing the scaffolding in the back alley, where neighbours were having their wall painted. I knew the scaffolding wasn't high enough to enable anyone to climb over my wall, if that was what he was trying to do, but I did not sleep well that night.

I was still wondering who it could have been in the morning. The ghost guardian, come to claim his money? The creep who'd threatened me the other day? Some bored teenager with nothing better to do than go scare the bejesus out of the weird foreign woman? There were no other foreigners in this part of the Medina; they usually bought property in the more touristy areas at the top of the two tala'as. The mystery was never solved.

A few days after this incident, my houseguests arrived. John and Nicole were travelling minstrels in a folk duo called Cloudstreet and were going from festival to festival in the United Kingdom during the summer. This was their first visit to Morocco, and was a good excuse to do some sightseeing myself.

Fez, like all Islamic cities, is centred around the souks. We went first to the food souk in R'Cif, where hundreds of tiny stalls are piled with vibrantly fresh vegetables, fish, meat, olives, coffee, spices, sweets. The meat stalls often display the heads of camels or goats, and I had got in the habit of going early whenever I bought meat, before the flies got to it. The practice of not refrigerating meat may sound unhygienic, but as it's usually killed and sold the same day it's considerably fresher than that found in Western supermarkets.

I couldn't bring myself to buy chickens, though – they were a bit too fresh. Looking a squawking chook in the eye while it was being weighed and then having its neck wrung at my behest was beyond me.

In the souk of the artisans' guilds, everything is made by hand, much as it has been for centuries. Seeing it through John and Nicole's eyes, I realised anew how much it was like being transported back to the Middle Ages, except that many of the goods produced now are exported or sold to tourists. In Place Sefferine, a constant rhythmic tapping sounds as coppersmiths make kettles, couscousiers and cooking pots. Some are for hire, big enough to whip up a feast for two hundred or so of your closest friends. Nearby are the knife makers, who, with a single foot operating a dangerously spinning stone grinder half their height, sharpen blades to surgical precision. In an adjoining street, brass makers cut, shape and emboss lanterns, plates and various household items.

In other ancient workshops, wrinkled cobblers sew bright yellow leather from Fez's famous tanneries into men's babouches. In the dars devoted to selling carpets to unwary tourists, looms are powered by teenagers and women, who need extraordinary patience to make the hundreds of knots required. Down in One-Armed Ali's weaving shop, the only sound is the clatter of the wooden shuttle flying back and forth over the warp. And in a little workshop in a side alley can be found the last surviving brocade makers in Fez, working on looms so complex they take days to thread and need two people to operate. The fabric they produce is sought after by designers in Casablanca and Rabat.

Countless other craftsmen service these artisans. The men who dye the yarns have permanently stained arms from hefting skeins out of steaming vats and wringing out the excess dye, before hanging them to dry in a rainbow of colours. A forge is constantly in action, where the blacksmith melds metal tools to order. The brass makers are serviced by speciality shops selling teapot spouts, handles and feet. Belt makers supply the shops selling long, mediaeval-style dresses for special occasions, and the hat sellers are close to the djellaba makers.

Some crafts are dying out as they become less sought after. The street of the saddle makers used to be full of workshops, but now only a few remain. And just a couple of gunsmiths are still licensed to make rifles for the Fantasia riders, who charge about on Arab stallions firing off volleys of shots in a display that is now mainly a ritual for tourists. But in the souk of the wedding-chair makers, fabulous thrones are still in big demand – plywood frames covered in glittering gold and silver fabric.

Every quarter in the Medina has workshops of carpenters, indispensable in a city whose skeleton is made of trees. Carpenters too have their specialties. There are those who repair and build houses, and others who decorate them, carving intricate designs for doors, tables, chairs and cupboards.

The henna souk at the bottom of the Tala'a Kbira is a quiet oasis, with a big plane tree shading a small square crowded with tiny shops selling pottery, pot-pourri, henna, argan-oil soap and rose moisturiser. At the back of this, housing another collection of shops, is an old building which was once an insane asylum. The sixteenth-century traveller and writer Leo Africanus, chiefly remembered for his *Description of Africa*, which became the basis of knowledge of Africa for scholars in the West for centuries, worked in the asylum for two years. He wrote that the mentally disturbed received no treatment other than being fed and fastened to the walls with iron chains. When they were brought food a whip was always handy to 'chastise those that offer to bite, strike, or play any mad part'. The hospital was still in use as late as 1944, although treatment methods had improved somewhat, and musicians occasionally came to play to the inmates.

For all the souks to function, the goods need to be moved around. Making your way through the streets of the Medina is a constant exercise in avoidance. You have to squeeze into doorways so that you're not mown down by heavily laden donkeys or mules, or wiry old porters with impossible loads. Fassis seem to have a special awareness of these ancient modes of transport and move instinctively out of the way, yet they are quick to lend a hand when a porter or hand-cart driver needs help. Tourists, on the other hand, are a real worry; they wander around in a daze, as oblivious to danger as a puppy on a highway.

Turning into the alley that runs past Karaouiyine University, I led the now footsore Nicole and John to my favourite Fez café. Liberace was waiting to welcome us, as he had done for patrons for more than forty years, resplendent as usual in a white suit with peacock feathers sprouting from one lapel and medallions on the other. His hair was hennaed bright red and teased into a fetching frizz. Large brown eyes dwarfed a mouth that was constantly in action.

His real name was Abdulatif, and although retired he still often turned up at the café. He had never married, and the staff and patrons of the place were his family. His other pastime was keeping his decrepit Renault 19 in immaculate condition. It was never driven – simply owning it gave him sufficient pleasure.

We squeezed up the café's spiral staircase, bending our heads as we arrived on the first floor, which is so low you feel like a giant in a dwarf world. I had dubbed the establishment Café Seven and a Half, as this oddity reminded me of the similarly height-challenged rooms in the film *Being John Malkovich*. Sipping rosewater milkshakes, we watched the constant flow of people and animals in the alley below, one of the busiest and narrowest in Fez.

Early next morning, we set out by grand taxi to the Roman ruins at Volubilis, a couple of hours' drive from Fez. Whereas petit taxis are restricted to intra-city fares, grand taxis travel between towns. The driver of our rickety old Mercedes was a grey-haired, tubby fellow of about sixty who won my respect when, shortly after picking us up, he stopped the cab and dashed out into the traffic to scoop up a kitten that was about to be flattened. Later he spent long hours waiting while we wandered around the ruins. We returned to the taxi to find him performing prostrations on a prayer mat rolled out in the dirt.

The World Heritage-listed Volubilis ruins sit on a rise overlooking a vast plain and cover an area of 4500 square metres. We arrived before the tour buses, but I was disturbed to see men in a field nearby digging what looked to be footings for the foundations of a large building. I found out later that a hotel was going up, but it was uncomfortably close to the ruins for my liking, and given their heritage status I couldn't understand why permission had been granted to build there.

Settled by the Romans in 40 AD, Volubilis was the breadbasket and administrative centre for the westernmost Roman province of Mauretania Tingitana. The daughter of Cleopatra and Mark Antony once ruled there with her husband, a Berber prince. When the Roman garrison withdrew at the end of the third century, many residents stayed on.

I had a vision of those Romans who had chosen to remain, sitting around in what had become a rural backwater, remembering the glory days of a vibrant and dynamic empire. The town was devastated by an earthquake a hundred years later, and although it was resettled by Latin-speaking Christians, it did not become significant again until the Arab invasion in the seventh century.

The foundations of the ancient stone houses are grouped around the remains of a Roman road. One of the larger houses is known as the House of Orpheus, named for the mosaic covering the floor of a reception room. Orpheus is surrounded by a variety of African animals, real and imagined. Once, the house also had a swimming pool, courtyard, garden, toilet, kitchen, and several other decent-sized rooms, all with underfloor heating. It was more appealing by far than many modern houses.

It's easy to see at Volubilis the origins of Fassi architecture. The Fassis still have courtyard houses, with *zellij* that bears more than a passing resemblance to Roman mosaics, although with geometric rather than figurative designs. Art in Islamic societies is generally used to display the underlying order and unity of nature, which is seen as a representation of the spiritual world. The bathhouses of Fez, its bakeries, guilds and public fountains, also owe much to the Roman style.

For public spaces, Volubilis had a large forum, civic buildings and a triumphal arch. Lining the main road, which is wide enough for two chariots to pass at speed and has a covered drain running down the middle, are the remains of shops and houses. Walking through the ruins it occurred to me how permanent the town must once have seemed to those who lived there, but after six hundred years or so, life took an unexpected turn.

In 683 the Arabs swept into North Africa on horseback. It was half a century after the death of the Prophet Muhammad and they were passionate about spreading the word of Islam. Despite an initial resistance, many Berbers found the certainty and absoluteness of Islam appealing, and were won over in a way they had not been by the Christianity of the Romans. Those Berbers who converted to Islam helped spread the religion south of the Atlas Mountains and north to a country they called Al-Andalous, now Spain. They made excellent soldiers (as the French were to find more than a millennium later), having honed their battle skills in intertribal warfare, and by 718 they had taken over most of the Iberian Peninsula.

Meanwhile the acceptance of Islam in Morocco was far from wholesale. Pockets of Christians remained, along with Berbers who still followed their indigenous religion. It wasn't until the arrival in 788 of a descendant of the Prophet Muhammad, a man who became known as Moulay Idriss I, that greater unity was achieved.

Idriss had fled Damascus for political reasons. He must have been a self-assured man with considerable charisma, because they made him King not long afterwards. Idriss became so popular in fact that the caliph of Baghdad felt threatened and sent his personal poisoner to get rid of him. After Idriss was buried, his wife, a young Berber girl, was discovered to be pregnant. She gave birth to a son, Moulay Idriss II. He was supposed to have been a remarkable child who could read at the age of four and recite the entire Koran by eight. At twelve he was appointed ruler.

Moulay Idriss II became known as the founder of Fez. While it had been his father's idea to build a new capital (having thought Volubilis a bit on the small side and too Roman in appearance), it was the son who was left to fulfil the vision. The name Fez means 'pickaxe' in Arabic, as this was the primary implement used to construct the city, and legend has it that a golden pickaxe was discovered while the building was being done.

John and Nicole left in mid-September and I had only one more week before I returned to Australia. I started looking in earnest for someone to oversee the restoration when Sandy and I returned the following year.

I had met an architect called Hamza, who'd done a wonderful job on his own house and overseen restorations for other people. An Iraqi refugee who'd studied in Europe, Hamza had a partner called Frida who was a graphic artist. Both spoke excellent English, and I invited them to dinner to discuss my plans, even though Hamza had told me he was far too busy to take on our house. But when he walked in and saw the open courtyard, the *massreiya* with the wonderful ceiling and plasterwork, and all the other interesting architectural details, he became enthused.

Dinner was beset by a few problems. Because the floor was at a strange angle, my new stove sloped forward and the fry pan would slide to the floor if not watched. I'd bought three turkey legs for dinner and put them in a supposedly heatproof dish at the bottom of the oven, with some vegetables on a higher shelf. I'd heard a loud crack shortly after turning on the gas, but it wasn't until smoke began pouring from the kitchen that I went to investigate. The dish had shattered and the turkey pieces were lying singed on the bottom of the oven. I rescued them and put them on top of the roast vegetables.

It didn't seem to matter too much, as I'd made a salad with fresh figs, mint and soft cheese, and a side dish of beans in tomato and onion. With a good bottle of Sahari Reserve to wash it down, it was very edible, and smoothed the way for further conversation about the house.

Over dinner I suggested a compromise to Hamza. What if he were to act as a supervising consultant, paid on commission, visiting once or twice a week to make sure things were on track? If his team wasn't available, I could hire another to do the work.

To my surprise he agreed, even suggesting that, to get a head start, he'd arrange to have the carpentry work done on the doors and windows immediately. I was thrilled.

The next day, Hamza returned with his carpenter, who had a look around and quoted twenty thousand dirhams, which covered two new doors, three windows, a set of traditional shutters and numerous repairs. It did not include cleaning the wood or removing the paint from the forged iron on the windows and catwalk.

As we trailed around in the carpenter's wake, Hamza told me what he thought would need doing next year – this *zellij* taken up, these walls stripped of plaster, this decorative plaster repaired. One of the things that needed immediate attention was the kitchen ceiling, which was about to cave in. On the floor above it was a tap without a drain outlet, and years of sluicing the floor had resulted in water soaking into the insulating layer of earth above the beams, adding to the weight and causing the wood to rot. I felt both excited and alarmed by the extent of the work he was suggesting, but having seen his own house, I trusted him.

Hamza also took me to see a house he was working on for a French man. The entire place had been stripped back to the bricks, and a team of men were removing a layer of rubble from the roof. In the stream of light from the hole they had made, a cloud of dust motes swirled. I felt a familiar tightening in my chest, the onset of an asthma attack, and quickly escaped to the street. I began to doubt that we'd be able to live in our house while work was being done.

And what must the dust be doing to the workers? The air was filled with minuscule lime and wood fragments, and who knew what else besides, none of which could be healthy. The only workers I'd seen wearing any protection were the men shovelling rubble on the terrace, and even then it was just a simple scarf. I reminded myself to bring some masks back with me, and to make sure ventilation was adequate.

At Hamza and Frida's house, I gave him a cheque for twenty thousand dirhams for the carpenter, along with a list of the work agreed to. In the back of my head was my grandmother's voice, telling me there were two types of bad payers, those who paid in advance and those who didn't pay at all. When I'd suggested paying after the carpentry had been done, Hamza said he didn't work that way. I consoled myself with the fact that, as he'd be in charge of our project, it could be considered a start-up fee.

I then spent a couple of hours taking photos of Hamza and Frida's place for their website, as a favour. They were planning to run it as a guesthouse, something it was ideally suited to, with its huge courtyard and terrace. One of the main suites was without doubt the most beautiful room I had seen in Fez. The view across the river had probably changed little since the place was built hundreds of years before, and the interior decoration was superb — intricate, detailed, with delicate paint and plaster work.

The other thing I needed to organise before I left was someone to stay in our riad. I certainly wasn't about to ask Larbi again, but David had a young American student who was willing to act as caretaker in return for a place to stay. Sarah was twenty-two, petite and pretty, with an eager-to-please manner. When I gave her a tour of the riad she loved everything about it, and she said all the right things — she was serious about her studies, and not a party person. She'd lived on her own a lot and didn't like large numbers of people around. That was the problem in her current student house, which was constantly full of people. A Moroccan friend had advised her not to invite the neighbours in because they gossiped about the valuables in the house, and word got around. I'd broken that rule, I thought ruefully.

There was just one thing that concerned me. Sarah was wearing a tight T-shirt with a plunging neckline and the slogan 'Boys R Toys'. Bringing her to the house, I had seen the way the local youths stared at her, as if they were about to eat her up. This district was one of the oldest in Fez and very traditional. Expectations of behaviour here were different from the more touristy areas and the Ville Nou-velle. Local women dressed conservatively, with no bare heads, legs or arms. Western women could get away with bare heads, but bare arms and close-fitting shirts were pushing the boundaries.

It was the equivalent, I explained to Sarah, of living in a Western city and seeing the woman next door heading out to the shops in just her bra – this was the way Fassis saw it. A local man had told me recently that only prostitutes had bare arms. As the new person on their patch, Sarah's behaviour would be watched closely. People would know very quickly where she was living and whether she came home late at night alone, so it would help, I pointed out, if she had the reputation of being respectable. That way she'd be less likely to run into trouble.

No doubt she thought me a stupid old bat, but I figured that since she was the one moving into traditional Fassi territory, it wasn't up to the locals to change their perceptions, but for her to be respectful of theirs. Moreover I didn't want her creating problems in my relations with the neighbours.

Months later, when I returned, I learned she'd taken my advice to heart and bought a couple of djellabas for wearing in the Medina at night.

On my last evening in Fez, I took a drink up onto the roof and gazed across the city, watching the horizon shift from gold to pink to deep blue. A dark stain of smoke from the potteries hovered over the north. I was far from alone – on nearby rooftops, other women and their cats were doing the same thing, except without the gin and tonic. As devout Muslims don't drink and the Fez Medina is a holy city, alcohol is not on public sale. To buy it you need to go to the Ville Nouvelle, and if on your return the taxi driver hears the telltale clink of bottles, he will most likely refuse to take you and your shopping. Many Fassis have a similar attitude towards the demon drink as we do to heroin.

Now that my departure was so close I didn't want to leave. Apart from seeing Sandy and friends, I wasn't looking forward to going home. Western cities may have their physical differences but the organisation of modern life, with its automated services, cars and franchised businesses, lends a similarity to them. Australian streets, by comparison to those of Fez, seem devoid of colour and life. It struck me that we have traded vivacity for the myth of safety. We exist within bubbles of cars and houses, and view the rest of the world through the glass wall of television. Where are all the people, donkeys, cats, the women waiting in doorways with trays of uncooked bread? The touts and spivs, the children playing in the alleyways? The evening streets crowded with those who hide all day from the sun? All the myriad small dramas that make up everyday Fassi life.

I had never been in such a vibrant, vitally alive place.

When Sandy and I returned to Fez at the beginning of the
following May his excitement was palpable. Mine was
more subdued. I was apprehensive about what might have happened
in our absence. What if the carpenter had removed the decoration
on the big doors to the salons? Or messed with the wonderful
massreiya ceiling?

We arrived to find the inevitable dust, along with workmen's
tools and boxes of garbage strewn about, but it was still good to
be back. The courtyard was filled with sunlight, there were oranges
hanging out of reach on one of the trees, and a couple of baby
sparrows were hopping around. Someone had put a grubby teddy
bear on the fountain spout, adding a surreal touch.

The carpenter had done part of his job. The salon doors had
been repaired, decoration intact, as had a couple of others, and
upstairs there was a new set of shutters and two new doors. But

some of the work looked slapdash. The bathroom door frame, which needed a section of rotting wood replaced, had had a blob of cement slapped onto it. There was a new window in the kitchen that I hadn't asked for, while the three I had listed remained undone. The smell of fresh cedar permeated the air, but it looked as if Sarah hadn't been here in weeks. There was off milk in the fridge, and the mattresses were stacked against the upstairs walls. When I got round to ringing Sarah later she said that the carpentry work had forced her to move out.

We'd barely deposited our luggage when there was a bang on the door. It was a young girl from across the alley with welcoming cups of tea. She pointed out her door as the one I knew as Khadija's. I had never seen the girl before and asked where Khadija was, but couldn't make myself understood. No matter, I thought, Khadija would be over the minute she heard I'd returned. In the meantime we busied ourselves with making the riad habitable.

Twenty-four hours after we arrived, there was still no sign of Khadija. I missed seeing cute little Ayoub playing in the street, and having Khadija pop her head out the door whenever I went out. Even though at times I found her exasperating, she had been part of the fabric of my life here and I'd been looking forward to seeing her. I'd bought her a new set of sheets, a painting set for Ayoub, and printed up a photo of them both. Eventually I asked another neighbour, who told me that Khadija and her family had moved to a small town in the countryside. I was surprised by how disappointed I felt on hearing this, and not only because she'd helped me so much.

Sandy and I were to be in Fez for seven months on this trip, and there was much to be done. We arranged to meet Hamza to discuss the next stage. As only half of the carpentry work had been completed, and that of variable quality, we were apprehensive about him overseeing the rest of the work.

He arrived late, well into the afternoon, with his Irish accountant in tow, and was his usual charismatic self, exuding a *bonhomie* and confidence that were belied by the endless cigarettes he smoked. He had fallen out with the carpenter who had given the original quote, he told us, and the replacement was more expensive. Hamza would return with the new carpenter to address the shoddily done work, and assess what more he could do for the money we had paid. Fair enough, I thought.

With the accountant he then had another look around the riad, muttering about the huge amount of work to be done. There were a couple of 'pregnant bellies', as Hamza put it, where moisture had entered the ceiling beams, forcing pressure downwards and making the walls bulge – what the engineer had called the *flambement*. We were a bit shocked when he said that the beautiful old *zellij* in the courtyard needed ripping up and relaying, as it was uneven. We thought the uneven patches, caused by tree roots, was part of its appeal.

We moved on to the issue of money. Hamza wanted forty per cent of the cost of the work up front, with each stage paid for in advance. It made sense from his point of view; he had been left holding the baby on one job and ended up bankrolling it out of his own pocket.

But forty per cent up front was a commitment we didn't feel comfortable making. What if a few weeks down the track the arrangement wasn't working? But our choice was limited – there were few people in Fez capable of supervising the work – and we parted saying we would get back to him in a couple of days.

Regardless of who oversaw the restoration, we wanted to get started as soon as possible. As there would be strangers working all through the house, we needed a place to store our valuables, so in time-honoured fashion we decided to buy a wooden chest. We found one we liked in an antique shop near the tanneries. It was embossed with brass shields and had a lock with a moveable pin mechanism that dropped in and out of place as the key was turned. Similar locks have been in found in Egyptian tombs dating back four thousand years, and are the forerunner of modern pin tumbler locks.

'It was made by the Tuarag people from the Sahara,' the shop-keeper said, oozing sincerity. But the price tag was far beyond what we'd planned on spending.

'I make you a special deal,' he whispered, as if afraid his regular tourist clientele would overhear, and more than halved the price. But it was still way above what we thought it was worth.

'Two thousand dirhams is what we can afford to spend,' I said. We stuck to that, and miraculously ended up buying it. A porter appeared and threw the heavy chest onto his back as though it were made of balsawood. He moved so fast through the crowded streets we were forced to trot to keep up with him. Jogging along in his wake, I noticed that the finish of the wood was uneven and there was a small chip showing blond wood beneath. The chest had probably been knocked up in Fez the previous week. I felt stupid for being diddled in the dimness of the shop. I did like the chest, but it was certainly only worth as much as we'd paid, if that. Expert traders, the Fassis had had hundreds of years of honing their skills on gullible foreigners like us.

The day before, I had gone to the souk to buy herbs. 'How much?' I asked an old man holding up two bunches of oregano.

'Ten dirhams.'

Huge bunches of mint, parsley or coriander were usually sold for a half a dirham, but I was too tired to argue. I took half of what he held out and gave him five dirhams. As I walked away he said something to the men nearby and they burst out laughing. I was sure it concerned my ignorance, but how could I blame him for trying to get the maximum from someone who could obviously afford it?

Despite such experiences, I tried to rid myself of the attitude that all Moroccans were out to cheat us. David had told us of an English woman who was buying a house in Fez, using a Fassi friend of his as a facilitator. She spent a lot of money engaging a London solicitor to organise power of attorney for him, and then discovered that this could only be done through the Moroccan consulate. David's friend should have known that, she maintained in a long letter of complaint to him. Moroccans needed to understand the standards Westerners expect, she had written, threatening to cancel her cheques and call the whole deal off.

If she was this upset at such an early stage in the process, we wondered how she'd go when she ran into real problems. She and Fez were clearly not meant for one another, Sandy observed.

We heard other stories about property-buying foreigners from David, and he introduced us to an Englishman who'd flown to Fez on a four-day visit. Although it was his first trip to Morocco, he'd bought a house that morning and was talking about quitting his job as maître d' in a trendy London bistro and moving to Fez full time. I wondered if he had any conception of what life would be like here. He might think he needed a break from the pubs and clubs of London, but living in the alcohol-free Medina, with few options for eating out, surrounded by conservative and curious Moroccans, might rapidly lose its lustre if he didn't have a genuine interest in the culture. We got the impression that at this stage it was all just colour and movement to him.

The Englishman told us that programs about buying houses in other countries were a national obsession in Britain. 'It's the British way of recolonising the world,' he said.

How true, I thought. Many houses in the Marrakesh Medina are now foreign-owned. A lot of the fly-in, fly-out expats limit their interaction with Moroccans to servants and shopkeepers — a replication of the colonial experience, and hardly the way to maintain a vibrant and cohesive community. Sandy and I were determined that, despite the cultural and linguistic barriers, we would not isolate ourselves within the expat community.

Right now that community was troubled by the fact that local authorities were cracking down on illegal guesthouses. The criteria for obtaining a licence were so strict — stipulating rooms of a certain size, ensuite bathrooms, televisions and refrigerators, air-conditioning, a salon and a minimum of five rentable rooms – that without purpose-built facilities it was difficult to fulfil them.

A number of foreigners had bought property in the Medina with the intention of staying and making a living letting rooms to tourists. Now, having done up their houses, they were jumping through hoops trying to meet the regulatory requirements. One woman had a beautifully restored house with five bedrooms, but because it was considered too small and the bathrooms were not ensuites she was refused a licence.

Dozens of illegal guesthouses had recently been closed down in Marrakesh. The introduction of cheap flights to that city had led to an explosion in the number of foreigners renting out rooms. Not all travellers were looking for a five-star hotel with equivalent prices, and because guesthouse regulations were so strict, some people ignored them and let rooms anyway. This meant they did not pay tax and their profits were sent out of the country. As cheap flights to Fez were about to start, the authorities here were determined not to let the same problem develop.

Frida and Hamza had been trying for months to get a licence for their magnificent dar, which had taken twenty people two years to restore. It seemed that there were three ways of getting any sort of authorisation in Morocco. You either conformed to the letter of the law, knew someone powerful, or gave 'presents' to officials. Out of desperation, with loans to service, Frida and Hamza had taken in a few paying guests regardless. The next thing they knew, the district official turned up in the company of their neighbour, who had made a complaint about them. They were handed a document ordering them to cease and desist, and had to arrange for their guests to go to hotels. Naturally they were very upset.

Frida and Hamza's neighbour was convinced that part of their courtyard belonged to him. Running through it was an easement allowing access to his carpet shop's rear door, which hadn't been opened for years. Now that Hamza and Frida had beautified the courtyard, he said he wanted to use the door again. Closing down the guesthouse appeared to be part of his strategy to make Frida and Hamza purchase his easement rights.

Hamza's response was to say that opening the door was a terrific idea: the neighbour's carpet clients would be able to see what a lovely place their dar was to stay in, and in turn Frida and Hamza could direct their clients to his shop. The down side of this proposal was that when the neighbour's back door was open, it revealed a scenic view of his toilets.

We finally received the deed to our house. Sandy and I met the scribe in a café, where he handed over a single sheet of paper in Arabic, to be joined to the two-metre scroll already in our possession. The only words we could discern on the scroll were our names, but with the help of someone who read Arabic I learnt that the scroll dated back to the beginning of the French protectorate in 1912, when a new system for property registration had come in. The house had been sold in 1932 by a group of people whose names took an entire page – probably an extended family who'd inherited it.

Since then, the riad had changed hands half a dozen times, with the longest period of ownership being the thirty years to 1977, when it was purchased by the couple from whom we'd bought it. The long list of names was a reminder that the concept of owning an ancient house is an illusion – we were all just passing through.

Sandy and I were still mulling over our decision about Hamza. From what we could gather, he was expensive and overcommitted. We knew an expat called Amanda whose restoration had been done by Hamza's team. Her house had been gutted and rebuilt, and while she was pleased with the work, we were horrified to hear she had so far spent more than three times the purchase price. And her house was tiny compared with ours. There was no way we could afford that kind of expense. Amanda had had moments when she wished she could get out of the arrangement, but she was already past the point of no return.

Warning lights were flashing in my head, but if we didn't engage Hamza, who else was there? I did know of a building contractor, but he was subcontracted to Hamza.

As luck would have it, we met a young English couple, Jon and Jenny, who had restored a dar in the Medina. It was medium-sized, of a much later period than our riad, but it was exquisite. There were multiple rooms with blue and white *zellij*, lots of detailed plasterwork, and the wooden shutters were in excellent condition. The work had been done on a reasonable budget by a team of workers they'd organised themselves.

An idea started to form. Why couldn't Sandy and I manage the restoration ourselves and hire the help we needed? We could engage Jon and Jenny to help us find tradespeople and for general advice. Although they'd only been in Fez a year, they had good contacts and spoke a smattering of Darija. True, they weren't nearly as experienced as Hamza, but he appeared to be so thinly stretched we doubted we'd be getting the full benefit of his experience. Managing the project would be a lot more work than we'd planned on, but I had a clear idea about what should be done.

The more we discussed it, the more possible it seemed. Sandy and I worked well as a team. Since I spoke the better French and had design skills, it made sense that I be the one to set each stage up and source the materials. Sandy, being empathetic, humorous, and possessed of vast reserves of patience, is an excellent people manager. His role would be to micro-manage the building work with the help of a translator. 'You can dream the dream and I'll manage the nightmare,' Sandy said.

And so it began. The first thing we had to do was get a building permit. This was called a *roqsa*, and as we discovered was the most important document we had to acquire. Without this three-month permit, nothing could happen. But getting a *roqsa* was no easy task and some people waited months to receive one.

I headed off to the *baladiya*, or local council office, where I found four women sitting at desks piled high with files. One was cleaning her nails, another was asleep. It didn't look as though the granting of permits was occupying their every waking moment. An obliging man who spoke English helped me fill out the necessary form in Arabic. I let him tick the boxes and then I signed it. I could have been agreeing to deed the house to him, for all I knew, but

I trusted him. No doubt these have been famous last words on numerous sorry occasions, but he seemed like a nice chap.

I hadn't bought a photocopy of my passport with me and had to walk back to get one. When I returned I lined up at another window to have the form authorised. Half a dozen Fassis were jostling for various permits: there were schoolgirls wanting authorisation for some unknown activity, a middle-aged woman whose identity card listed her profession as embroiderer, a man in a djellaba who looked old enough to have lived through the French occupation and retreat. When it was my turn the woman behind the counter took a break from serving me to help several other people. At last it was done. I lodged the forms with the nice chap, who told me an inspector would arrive the following day.

True to his word, there was a knock on the door after break-fast the next morning. A dapper man with a grey moustache introduced himself in excellent French as the chief inspector. I ushered him in and he proceeded to check out every room in great detail. When I showed him the *massreiya* he shook his head and tut-tutted about the state of the beams. Up on the terrace, he gazed out over the city, seeming to forget he was doing something as mundane as a house inspection.

'This area is one of the oldest in Fez,' he said. 'The émigrés from Tunis settled here in the ninth century.' Sensing my interest, he continued. 'When the refugees came from Spain they settled on the opposite bank of the Oued Fez, over there.' He swept his hand across the Andalusian quarter on the other side of R'Cif.

'During the eight hundred years we held Al-Andalous, the Moorish artisans developed their building skills to a high degree. That is why we have so many marvellous buildings.'

I'd heard this several times before. What I really wanted to know was why the artisans' skills had become so highly developed. Later, I found out. When Berber and Arab forces had taken over the Iberian peninsula at the beginning of the eighth century, they kept going up the European continent until they were stopped at Tours. They then turned their attention to consolidating their rule in Al-Andalous.

The Moors were unusual rulers in that they allowed the Christians and Jews to practise their own religions, which led to an increasingly complex society. Such tolerance was necessary, as there were several thousand Moors ruling over millions of Christians, and wholesale conversion to Islam would have been impossible. Instead the Moors imposed additional taxes and restrictions on non-Muslims, making it necessary to convert to Islam if you wanted to get ahead. By the eleventh century, Muslims outnumbered Christians in Al-Andalous.

The Muslim rulers faced a problem common to conquerers throughout the ages. Taking over a country was the easy part. Staying in power and running the place was extremely hard. Over the centuries, Al-Andalous broke down into more than twenty principalities, whose rulers competed to be the richest and most architecturally and culturally impressive. Each fought to recruit the most skilled artisans, poets and scholars, leading to an extraordinary flourishing of the arts and sciences, which had a flow-on effect to a moribund mediaeval Europe. The most famous Moorish legacy is Granada's beautiful Alhambra Palace.

Meanwhile the Christian rulers in the north gained in strength and picked off the principalities one by one. As each was overrun, waves of refugees fled to Morocco, among them leading artisans, who became the catalyst for a flurry of new building. Their rich artistic legacy can be seen in the exquisite plasterwork, carpentry and *zellij* of buildings like the Attarine and Sahrij medersas – the most beautiful of the ancient theological colleges in the Medina – as well as many riads and dars built by wealthy merchants, including our own.

The inspector grilled me about changes we intended to make to the riad, then declared that a second, specialist inspector would have to look the place over. And we'd also be needing a supervising architect. Another layer of bureaucracy and expense.

Getting a *roqsa* might have been complex, but arranging a telephone and Internet connection proved no easier. Actually, it seemed remarkable that such a thing was even possible, given that many people in the Medina could not afford water in their homes, but had to go daily to the public fountains.

At least when I arrived at the Maroc Telecom office the doors were open and there were no other customers. Then it registered that there was no one behind the counters. A security guard waddled up and told me to return on Monday. I was nonplussed. The opening hours were given as Monday to Thursday, eight-thirty to four-thirty, with a two-hour break for lunch. It was nine-thirty on a Thursday. What was going on?

Puzzling over this, I walked down to the Café Firdous, where I found David having his morning coffee.

'Well, it's Thursday,' he said when I told him what happened. 'Maybe they're getting ready for Friday.'

His phone rang. 'My next door neighbour's builder is removing all the old *medluk* from their outside wall,' he told me after the call ended, looking distressed.

Particular to Fez, *medluk* is a mix of sand and lime used as an exterior finish.

'I said I'd pay for it to be restored,' David continued. 'I thought we'd agreed, but the builder is taking it all off and replacing it with cement. He says if I want to pay for *medluk*, that's my business. The cement can be removed later and new *medluk* put on.'

I wondered whether the builder was hoping to be paid twice for the same job.

Salim, the engineer who'd inspected our house, was sitting at a nearby table. David called him over and Salim agreed to go and see if he could stop the work. According to David, there were issues like this every day. Old surfaces of buildings were removed, and replaced with new and inappropriate materials. It didn't help that contractors engaged by the government were paid by the metre, so it was in their interest to remove and replace as much as possible. In a strange way, the increasing wealth of the city was becoming a threat to the survival of the very aspects that made it unique, and therefore historically priceless.

I wished David luck and returned home to meet with a prospective cleaner, someone to mop the floors once or twice a week and perhaps do some laundry. While not averse to doing a bit of the old spit and polish ourselves, Sandy and I needed to concentrate on our building and writing projects.

I had asked around for a reliable and trustworthy person and was put in touch with a teenager called Damia, who brought along her English-speaking boyfriend to translate. As I gave them the guided tour, I stressed that I needed the floor washed with very little or no water, pointing out the bulge in the kitchen ceiling caused by water being liberally strewn across the floor above.

We sat at the table and a vigorous discussion ensued between Damia and her boyfriend, who finally said, 'Damia thinks it's a lot of work and she needs someone to help her.'

I was surprised at this. Khadija had never had any problems on her own. 'If she doesn't want to do the job, that's all right,' I said, and reluctantly Damia agreed to do it alone.

When she turned up the next morning I explained again the need to minimise water on the upstairs floor, pointing once more at the kitchen wall and ceiling. She disappeared upstairs, and moments later I heard the sound of water being sloshed over the floor. I rushed up and she looked at me in bemusement while I ran around in circles trying unsuccessfully to stop the water getting through the crack in the floor that led to the ceiling. When the flood had been stemmed I took the mop and mimed squeezing it out, saying shrilly, '*Petit l'eau.*'

Back downstairs, I resumed writing, then was amazed to hear

the slosh of water once more. I raced up the stairs two at a time, to see a large puddle disappearing through the crack. Perhaps my previous reaction had been so amusing she wanted a replay. I obliged, struggling to get rid of the water while she stood and watched me. Then I took her back to the kitchen and pointed yet again to where the ceiling was about to collapse. '*Pas avec l'eau!*' I yelled.

Finally she seemed to get it, and the rest of the cleaning passed without incident – in fact she did an exceptionally through job. The only cause for concern was the sound of breaking glass. When I went up to look later I couldn't find what had caused it. No doubt the offending object had been whipped out of sight.

The following week, I had a call from the guesthouse owner who'd recommended Damia. She wanted to know if I was happy with her and I said I was, with the exception of the water incident. It transpired that this woman had just lost a large sum of money. She'd been packing to go away and had momentarily left her money belt on her bed. Damia and her boyfriend were there at the time, along with her cook. Several hundred dirhams were gone, along with some foreign currency.

The latter had turned up in an unexpected way the following day. 'I'm sure you must have made a mistake and the money is here somewhere,' Damia's boyfriend said and proceeded straight to a guidebook, inside which, lo and behold, were the foreign notes. He denied there'd been any theft.

'It's as though the culprit felt guilty and wanted to return some of what they took,' said my friend, who'd decided not to go to the police but was weighing up whether to sack all three.

'I doubt Damia's boyfriend is the thief, but he's probably trying to cover for whoever is.'

I decided that until the situation was sorted out I wouldn't be employing Damia any more. The telephone conversation shifted to how difficult it was to get good help, and I hung up with the ghost of Somerset Maugham whispering in my ear: 'I made up my mind long ago that life was too short to do anything for myself that I could pay others to do for me.' Was I, despite my best intentions, succumbing to a colonial way of life after all?

It was now the middle of May. Spring had well and truly arrived and the air was redolent with the scent of roses: you could buy kilos of petals at the souk, from which many people distilled their own rosewater for use in cooking and ceremonies. We were asked to dinner by an American couple we'd met. They had spent all day having a cooking lesson from a Berber chef, going around the souks and selecting the best and freshest produce, then being galley slaves while he created the magnificent meal we were invited to share.

The first course was fresh goat's cheese with herbs, presented with a radiating sun of roasted red capsicum, followed by an eggplant and chilli salad, then a mouth-wateringly fresh chicken dish with prunes, roasted almonds and potatoes. A melon and mint dessert followed.

I sat next to a young Fassi woman of unusual beauty called Ayisha. She wore a dark-red Saudi Arabian tunic and trousers, and had such perfect sculptural arches over her eyes that I had a pang of eyebrow envy. Her English was excellent – she was studying it at university – although she had an American accent from watching soap operas.

Ayisha was twenty-three and from a desperately poor family. She was longing to break free of her confined existence.

'But I don't want to marry a Moroccan man,' she told me. 'They don't treat women like equals.'

A couple of years previously she had been *affianced*, as she put it, to a fellow who'd asked her father for her hand. To her father this meant she could forget about studying, so she rebelled against the proposal and insisted on her right to go to university. Her father was unimpressed and their relationship was now strained.

Despite thumbing her nose at her father and pursuing an education, there was one area where Ayisha wasn't prepared to break with tradition. She was a virgin, she told me, and would remain so until she married. I enjoyed her lively mind as we ranged over everything from the role of women to religion and vegetarianism. Ayisha did not eat meat, as she 'empathised with animals acutely', a considerable attitude shift in a culture that valued animals largely for their utilitarian purposes.

In lots of ways Ayisha typified the young Moroccan women who were causing a seismic upheaval in their extremely traditional culture. This had resulted in thirty-five women being elected to the Moroccan parliament in 2003. And in 2006, fifty women had graduated as religious leaders, the first contemporary female group to be officially trained as such in the Arab world. They could do everything the male imams could, except lead Friday prayers in a mosque. This was an unheard of prospect just a few short years before.

Morocco is now a leader in the Muslim world for female rights and freedoms. The year 2003 also saw the introduction of a new family law code, known as the *Mudwana*, an historic piece of legislation that allows women to press charges against their husbands for domestic violence. Before that, a wife needed a witness to such acts before she could lay charges – an impossible situation. Under the *Mudwana*, women also have equal property rights and custody rights, and no longer have a legal requirement to obey their husbands. Forced marriages are illegal, and polygamy is permitted only if a judge can be convinced that both wives will receive equal treatment. Men cannot divorce their wives just by ritually saying 'I divorce thee' three times, as they could before 2003, and they also now have a legal responsibility for any children born outside marriage.

When the *Mudwana* was introduced thousands marched in support, but tens of thousands of fundamentalists protested, seeing increased freedom for women as a challenge to Koranic teachings, and a slide towards the immorality perceived in Western societies. The King and the government admirably maintained their resolve to ratify the law, although its passage did not change everything overnight: many Moroccan women are illiterate and don't know their rights, or are afraid to use them.

The simultaneous introduction of women into parliament will hopefully ensure that these reforms are not eroded, but the increasing power of the fundamentalist Wahhabists across the Middle East and other parts of Africa is an ever present threat. This movement generally believes that women should be submissive and ought not have a role in public life.

'I think we're going to be great friends,' Ayisha said at the end of the evening, squeezing my hand. I hoped so too, and indeed, a few days later she came to visit at Riad Zany.

I showed her around, and when we entered the tiny room off the stairs from the catwalk I wondered aloud whether it had been a place for prayers.

'It's too small for prayers,' Ayisha said. 'You couldn't stand to do prostrations. Of course, in the old days there would have been slaves in this house. This was probably where they slept.'

I'd never thought about slaves living in our house, but it made sense. Looking around the room, I saw that it would fit two sleeping people. A third could have slept on the narrow, tiled mezzanine, which I'd assumed was for storage – but it was long enough for someone to lie on, with a niche on one side to hold a candle and a few clothes.

I looked up at the intricately painted ceiling running along the length of the passageway, illuminated by a couple of arched windows onto the courtyard. Its repeating geometric pattern would have taken ages to create. Had a talented slave done it in their spare time to make their situation more bearable? Or had a kindly owner paid for it to be done, to beautify an otherwise utilitarian space? It would have to remain one of the many mysteries of the house.

Ayisha told me that until the early 1940s every self-respecting

Fassi family owned a slave. They bought and sold them on the market like sheep.

'My great-grandparents had a slave,' she said. 'My great-grandmother told me that in the old days children were frequently snatched off the streets and sold. One young girl from the powerful Alouite family was out playing in the street when she was kidnapped.' Ayisha made a snatching gesture, her eyes wide.

'The girl's father loved her so much, he spent years searching for her. He even dressed up as a beggar and knocked on the doors of houses all over the city to find her. Finally a young woman answered who had an unusually shaped birthmark on the side of her face. He looked into her eyes and realised he had found his daughter.'

I wondered if the father had told his daughter who he was immediately, and whether she believed him, but I didn't want to interrupt Ayisha's story.

'Then the father went to King Mohammed V,' she continued. Mohammed V had ruled during the 1930s and 1940s. 'He told the King that one of his family had been stolen and sold into slavery, and the King said, "That is enough. From this day I am going to make sure this barbaric practice is outlawed."'

My interest piqued, I did some research on the subject, then engaged a local guide to take me to the sites of a couple of old slave markets. One of them was hidden away down back alleys near the mediaeval Muslim college of the Attarine Medersa. Following the guide's lead, I ducked through a doorway and found myself in a large square surrounded by high walls. There was a smaller adjoining square, enclosed by a raised platform and pillars and resembling a Roman forum. It seemed likely this would have been where the slaves were shackled for display.

I'd read that the majority of Moroccan slaves came from West Africa, brought over the Sahara by traders. On their hellish trek through the desert, they were made to walk on the outside of the caravan to protect the goods within from attack by nomads. Anyone holding up progress was killed. The mortality rate was staggering – up to eighty per cent of those taken died on the way.

I was astounded to learn that in addition to the slaves from West Africa more than a million Europeans were taken. These were captured by Corsair pirates between 1530 and 1780, in numerous raids that depopulated coastal towns from Cornwall to Sicily. In the summer of 1625 alone, more than a thousand unfortunates were taken from the west coast of England. Corsair pirates were renegade groups of Moors who'd been expelled by the Spanish. They sold their captives to work as labourers, galley slaves and concubines.

Imagine going about your business as a Celtic villager and suddenly being whisked off to a completely alien society, unable to communicate with anyone. Given the knowledge of the day, it would have been the modern equivalent of being captured by extraterrestrials.

'Slavery in Morocco was not like elsewhere,' said my guide. 'Slaves were treated as part of the family.' Seeing my sceptical expression, he added, 'They had a good life. Why would you mistreat someone who could take it out on your children?'

I raised my eyebrows. If it was such a good life they were going to, why was it necessary to take them by force? On the contrary, I imagined the slave market was an extremely sad affair, with friends and family who had survived the dreadful trek together being forcibly split up. As far as I could see, the only positive thing to come out of slavery in Morocco was the rich legacy of *gnawa* music, a fusion of African and Arab influences.

The slave markets were still operating at the time of the French occupation in 1912, when they were officially outlawed but in reality only driven underground. The French turned a blind eye to the practice among powerful families and some of their own countrymen, and in the meantime supply dwindled due to the tightening of national borders.

Now the small square was filled with a cheerful jostle of women bartering second-hand clothes, everything from baby bootees to elaborately embroidered evening dresses. There was a great deal of chatter and laughter and it was as much a social occasion as a market, a far cry from the atmosphere the place would have had when slaves were sold here. In the afternoons it was the men's turn, and they came to buy and sell leather hides.

I was perturbed to learn during later conversations with Moroccans and expats that a form of domestic slavery still exists in Morocco. Some wealthy families 'obtain' a young girl from the mountains, who is brought to the house as a domestic help and must do everything from cooking and cleaning to childminding. Human Rights Watch reports that 'girls as young as five work 100 or more hours per week, without rest breaks or days off, for as little as six and a half Moroccan dirhams (about 70 US cents) a day'. It's a difficult thing to police because the girls often have no identity card and so officially don't exist. If the head of the family where she works is questioned he simply claims the girl is a niece. Some cases of physical and psychological abuse have become public, resulting in outcry.

In early 2007, the government launched a program called *Inqad*, meaning 'rescue' in Darija, as part of its ten-year National Action Plan for Childhood. The program aims to eradicate the market that deprives little girls of any semblance of education or opportunity. Whether they can achieve this remains to be seen, but it's a move in the right direction.

The Moroccan government is also attempting to combat the trafficking of men, women and children to Europe and the Middle East for forced labour and sexual exploitation. While many of those trafficked are Moroccan, others from sub-Saharan Africa and Asia transit through Morocco, and some end up staying in cities like Tangier and Casablanca.

These problems are by no means unique to Morocco. On the other side of the equation, some crimes we take for granted in the West, such as those related to hard drugs, are relatively rare here. The loneliness and alienation that is often the reason for people in wealthy countries turning to drugs does not appear to be as much of a problem in Morocco; people are too busy coping day to day. And the fabric of community is as intricately inter-locked as the houses themselves; everyone knows what's going on with everyone else. There are no instances of people dying and not being discovered for months, or sometimes years, as happens in Western countries. When I was in Fez by myself the neighbours would regularly bang on the door to check that I was all right. It was a bizarre notion to them that I might want to be alone.

Shortly after my visit to the old slave market, I asked a neighbour if he knew anything about the history of our house.

'Oh yes,' he smiled. 'When my mother was a child a man called Bennis was the owner. He belonged to one of the wealthiest families in Fez and was a silk trader who often travelled to India. He had three wives.'

My neighbour held up his fingers, nodding at my quizzical expression. 'Yes, three. He used to live in a very rich house in another area with the first and second wives. There was a beautiful young woman, a slave from Sudan, working in that house. Bennis paid for her with camels and coral. He decided to make her his third wife, although he was getting old by then and she was very young. To set her up, he bought your house and gave it to her. Neither of the other wives knew about the third wife.'

So Bennis's third wife had gone from being a Sudanese village girl to a slave, to mistress of her own house – our riad. Quite a transition in one lifetime. I wondered how Bennis had explained his repeated absences to his other wives. Or maybe he was away on business so often it wasn't a big deal. But surely someone had seen him coming and going. Fez wasn't that large, and like I said, everyone knew everyone else's business.

'Did they have any children?' I asked.

'Two daughters. But Bennis disappeared when they were

young. He went to India when he was seventy and didn't come back. Some people said he found another wife there, others said he died of an illness. And there were those who said his ship sank.'

I imagined the beautiful Sudanese woman bringing up her two daughters and waiting and wondering why her husband didn't come home, until one day she realised he wasn't going to.

'So what happened to her after he disappeared?'

'My mother said she stayed for a while, living quietly, and then one day she was gone. She and her daughters moved to Casablanca to make a new life for themselves.'

Thereafter, as I went about my daily chores in the riad, I thought about that former slave girl. I wished I could bring her across the passage of time for a short while, to show her what we were doing to preserve her beautiful house. I had the feeling she would approve.

Some time later, I found out the real reason Ayisha disliked her father. When she was a child, she told me, she witnessed him beating her mother.

'And he and my brothers also beat my older sister,' she said. 'To escape, she accepted the first offer of marriage that came along. That was a big mistake. Now she lives in the mountains with her husband's family and they treat her very badly. She must wear a full veil and is virtually a prisoner there.'

There'd also been a couple of attempts by the husband's family to poison her sister, Ayisha told me. This reminded me of India, where they burn brides, usually because the husband or mother-in-law thinks the girl's family hasn't provided a sufficient dowry, or they have fallen behind with their dowry payments. Burning her can be disguised as a domestic accident and the husband can marry again without dishonouring himself.

'I was determined not to be weak like my mother,' Ayisha continued. 'I told my brothers and my father that if they laid one hand on me I would go to the police, and so they are wary of me.'

It took a particular determination to resist the weight of tradition working against Ayisha. And being some years younger than her sister, she had grown up with access to mobile phones and email, things that make a significant difference to the lives of Moroccan girls. While many are forbidden to have relationships with men before marriage, they can now communicate without their fathers or brothers knowing.

Ayisha confessed to a long-distance relationship she'd been having with an Englishman she met the previous summer. She had only spent a few days with him but he had asked her to marry him. She was thrilled about this, seeing it as a means of escape from her limited life and prospects.

My antenna went up, and I told her the story of an American friend of ours who had been working in London and on holiday in Morocco when he met a beautiful young woman on a beach. They had only a few words of French in common but that didn't stop them becoming madly infatuated with each other. A few months later, they married in a traditional ceremony with white horses, silver thrones, and the blood-spotted sheet to prove her virginity.

After the honeymoon, our friend took his bride back to London. I could imagine her arriving in the strange city full of hope for her new life with her kind and intelligent husband. The following year, she had a son. Whereas in Morocco she would have had the support of the women in her family, here she had no one. Her husband's family were in Los Angeles and he worked long hours, leaving her alone with the baby all day in a small flat in an unfriendly city. She went slowly crazy.

Eventually she hooked up with a local junkie and abandoned her husband. For two years he fought through the courts for custody of his son. Then one morning his wife arrived on the doorstep and handed him the child. Her boyfriend did not want him and she was unable to care for him any longer.

Our friend moved back to Los Angeles with his son. When he remarried a couple of years later, his new wife and the boy did not get on, and in his early teens the boy ran away from home, joined a gang and roamed the streets, constantly in trouble with the law.

Ayisha was intrigued by this story but could not see how it might apply to her. 'My English is good,' she said. 'And I know how to make friends.'

I tried to get across to her that England was not all light, colour and warmth like Morocco, quite the opposite. It could be cold, grey and expensive, and social acceptance was often difficult for immigrants. Because of their skittishness about terrorism, many English tended to view Muslims with suspicion. Ayisha was dreaming of a London that existed only in her head.

8

At the end of May, we found a supervising architect, as required for the building permit. Rachid Haloui was one of the best architects in Morocco. Based in Fez, he designed projects all over the country, gave lectures on Islamic architecture in France, and had written a book on the coastal city of Essaouira. He cared deeply about the Medina and had been the founding president of the Fez Preservation Society.

When Rachid walked into our courtyard and looked up at the big wall dominating one side he muttered, 'Interesting, interesting.' He observed everything else in the riad with little comment until he came to the pillars in the *massreiya*, two of which bore elegant script. Rachid, with an eye for detail, was immediately taken with these but had trouble reading the ancient Kufic letters, lovingly created in *zellij*. He pulled out a camera and took a photo.

We later discovered that they were phrases from the Koran.

One translated as 'If you follow Allah he will give you what you need', and the other 'Allah will take care of you'. It was not an Islamic scholar who translated them for us, but our devout and educated builder.

Up on the terrace, Rachid pointed out that our rear wall was in line with the brick fortifications that had been the city's outer walls at the beginning of the seventeenth century. This, he explained, showed that when our riad was built it would have been just inside the city's boundaries. Over subsequent centuries, houses had spilled out well beyond these limits so that our place was now situated in the heart of the Medina.

We had a good feeling about Rachid. As well as being highly experienced, he was a kindly, careworn man, honest and direct, with a genuine love of traditional architecture. The one disappointment was that he wasn't able to recommend a builder. He was used to working on a much larger scale, with teams of contractors. I guessed too that if we found our own builder and he performed badly, it wouldn't reflect on Rachid. They all had problems, he said, and added ominously, 'They must be controlled.'

But he did suggest an engineer to oversee the structural work, and also agreed to give us an attestation – a formal notification that he was in charge of the project – so we could get a *roqsa* and begin work. As soon as we found someone to actually do it.

I quizzed David about the builder who'd visited the riad the previous August and promised a quote that never eventuated. He was a contractor, David told me, and would simply employ another builder and take a cut.

The list of people who would need to be paid was multiplying like relatives at the reading of a will. We'd also had to hire another translator, to communicate with the workers and help me buy materials. Si Mohamed had spent a couple of weeks working for Jon and Jenny, and although inexperienced he seemed keen to learn. So we were already up for a fair whack before we even got to the builder, his labourers, the plumbers and electricians. Not to mention whatever carpentry work was still outstanding after we resolved the situation with Hamza.

David knew of an older, well-respected builder who used to be head of the mason's guild in Fez. In this role he had overseen the quality control for all the masons in the city, and would know which ones were capable of doing our job. Jon and Jenny were friends of this man's son, so I gave them a call.

Jon immediately saw a problem. 'Hmm, if I ring the son then he'll want to do the job.'

'But he already has a full-time job.' I knew the son worked as another expat's builder.

'That won't make any difference. Even if he can't do it himself he'll still want to control your builders and make money from you.'

So we were back with the problem of how to avoid hiring a contractor when what we needed was a traditional builder, to whom we'd be happy to pay above-average daily rates. Jon and Jenny didn't think the one they'd used was up to the complex structural work our house required.

I explained our predicament to Salim when I ran into him

one day on my way back from the souk. He suggested a contractor friend of his.

'We just want a traditional builder, not a contractor,' I said. 'We've engaged Rachid Haloui to oversee everything.'

'But you will pay much more,' said Samir. 'Three hundred dirhams a metre instead of one hundred. And why do you need an architect? You only need an engineer.'

I explained that we needed a supervising architect in order to get a *roqsa*, but Salim claimed this wasn't necessary; an engineer was acceptable. This was news to me but it was too late now. We had already appointed Rachid and his engineer. However, I was determined to avoid adding a contractor to the mix.

'Our friends did their restoration without a contractor,' I said. 'They got an architect then engaged the individual craftsmen. It was much cheaper.'

'How much?' Salim challenged, aware that any arrangement he had envisaged with his contractor friend was now vanishing.

I fudged on this. 'So do you know of any good traditional builders?' I pressed instead.

Reluctantly he admitted he did, and named someone, adding that some of his work was not far from where we were.

I followed him through several alleys until he pointed out a rebuilt exterior wall. It wasn't much to go on, but the work looked competent. Would he be capable of replacing our huge beams without the walls falling into the street? Only Allah would know.

'Will you give me this builder's phone number?' I asked.

'You can find him through me,' replied Salim.

I took this to mean that if I wanted the builder I'd need to engage Salim as the engineer, but I didn't need another one. 'Does he work on his own?' I persisted. 'Do you have his number?'

'Tomorrow I will ring you, *inshallah*,' Salim said.

Naturally he didn't ring, and I spent several fruitless days chasing him.

Sandy, meanwhile, had been working on a new novel and on our weblog, 'The View from Fez', which we'd started for friends. It had taken on a life of its own and was now the top Moroccan weblog in English, and needed constant feeding. Having been cooped up inside during the day, he relished our walks through the lively Medina of an evening.

In the streets around Tala'a Sghira, waves of Arabic pop music competed with Bollywood hits, jazz, and the roar of football. The scent of rose petals and incense and ripe fruit hung in the air, along with the smell of hashish wafting from groups of likely lads (making them far more docile than their alcohol-fuelled counterparts in Western cities). Snail sellers offered bowls of snail soup, and kids were fishing the tiny molluscs out with safety pins. Whole families were out for an evening stroll – children riding on their fathers' shoulders, boys playing vigorous games of soccer, people stopping to greet one another and chat in the warm night air. This was what living in a city without cars meant.

Our battle for a *roqsa* was not yet over. When I returned to the *baladiya*, armed with the document Rachid had prepared, the nice chap told me he'd sent the second inspector to our house but we weren't there. I was surprised. Either Sandy or I had been home during working hours every day, and I said as much as I gave him the paper.

He glanced at it, raising his eyebrows when he saw Rachid Haloui's name. 'You need to write a letter to your architect,' he said.

'Why? I can phone him. Anyway, he'll be at my house next Tuesday.'

'No, you must write a letter telling him that work will start. If you come back in an hour I will give you one.' He turned his attention to the next customer.

I left, puzzling over why I needed to write Rachid a letter when he already knew about the work. To fill in the hour I went and pestered Maroc Telecom. I had been assured that technicians would arrive the day before to install the phone, and of course they hadn't. The pleasant man behind the desk made a call, failed to get a satisfactory response, and took my mobile number instead, saying he would call when the technician was on his way.

To my surprise, a week or so later he did just that, and a technician turned up at the riad. He neatly strung a line through the kitchen window, up underneath the catwalk and in through the window of the salon. When they'd gone they rang twice to check that the number was working, and to tell us that perhaps as soon as tomorrow we would be able to make calls out, *inshallah*.

It was a simple but miraculous thing. This was the first time the riad had had a telephone in its 300-year-plus history. Decades were being leapt in a single bound. Next week the Internet was also due to be connected, *inshallah*.

Meanwhile, back at the *baladiya*, the letter was not ready, but no matter, the nice chap took a sheet of paper and proceeded to draft a letter in French telling Rachid that work was about to commence and that we would be honoured by his presence. I was then informed that I must pay a 250-dirham administration fee at another counter, then post the letter by registered mail and return to the nice chap with both receipts. Welcome to Moroccan bureaucracy. It all seemed arcane, but I had no choice but to comply.

I took the payment to a man in a side office, then trudged to the post office in sweltering heat. I sent the letter by registered mail as instructed, and hoping that the *baladiya* hadn't closed for lunch, I trudged back again.

The nice chap was nowhere to be seen. I presumed he'd escaped for lunch, and eyed the four women behind the counter who were sitting chatting and studiously ignoring me. After a couple of minutes, another fellow who was waiting told me the nice chap wasn't coming back. 'He got a phone call from his wife,' he said. 'She told him both his parents just died.'

I was astonished and dismayed. While I'd been worrying about relatively petty concerns, his life had altered inexorably. I mentally wished him well and vowed to be more patient with officialdom.

It turned out that the nice chap had completed my *roqsa* before he left. I had a strange eureka moment when one of the women finally came over to give it to me – a tiny, insignificant-looking piece of paper that allowed us to start work on the riad.

It was at once a huge relief and, because of the death of the nice chap's parents, a sharp reminder of how quickly life can change.

⌣

I finally got hold of Salim and managed to arrange a meeting with the builder he'd recommended. We met at Jon and Jenny's place, which was near where he was working, and he turned out to be about seventy, with thick-lensed glasses and only one eye. Wearing an immaculate cream djellaba and a white skullcap, he sat on the edge of his chair looking distinctly uncomfortable.

I found it hard to picture him scrambling up and down ladders and hefting heavy beams, or judging the straightness of a brick wall. On the plus side, I told myself, he was bound to have a vast wealth of experience and would be more reliable than younger workers.

Jenny conducted the interview through a translator, with the occasional question from me. I wondered if this venerable fellow minded being quizzed by Western women dressed in immodest clothing that revealed their ankles and hair. Perhaps not, because he agreed to return the following morning and take us to see the house he had almost completed.

But next morning the appointed time came and went. When the translator called him, the builder said he would come that afternoon but refused to be pinned to a precise time. I was getting the impression he didn't really want to work with us but did not wish to say so.

'This has been the problem trying to employ people all along,' Jenny sighed. 'They'll give a meeting time and then not show up, even when we call them half an hour before to confirm. It's the Moroccan unwillingness to disappoint.'

It seemed to me that Moroccans had their priorities sorted. Life is more important than work, and if something more interesting comes along, why not pursue that instead? This is why *inshallah* is such a useful word. If Allah wills me to sit here drinking coffee with friends rather than meet with a prospective employer, then that's the way it is.

A few days later, Jon had a lead on another builder. 'At least this one is under forty and has two eyes that both seem to work,' he texted me.

Omar came to see the house the following Monday. Big, brash and confident, he gave lengthy explanations via Si Mohamed of what was wrong and how he intended to fix it. It got a bit worrying up on the roof when he blithely announced that the entire job would take three weeks.

Jenny and I raised our eyebrows. 'We want a careful job done,' I emphasised.

'Oh, all right, two weeks then,' he replied.

Either Omar was Superman or he planned on bringing fifty sets of hands in. Or perhaps he was saying whatever he thought would get him the job. Whatever, he was at the top of our list of possible builders, although the only other name on the list was a one-eyed seventy-year-old who hadn't shown up.

Omar took us to see another house he was working on,

a project with which an Australian man we knew had been involved for a while, before he fell out with the owner. I rang him to get the low-down.

'Omar will be trouble for you guys,' he said.

It seemed Omar was a good builder but he could slip up through inattention and he was money-hungry. What builder wasn't? I wondered. The price Omar had quoted us seemed ridiculously cheap, but we planned to pay a day rate so he couldn't rush the job.

Our friend had some good advice: we should also pay a monthly bonus to keep Omar on track and prevent him going off to someone else's better paying job in the interim. We'd need to buy the tools for his team ourselves, and most of these would go walking if we didn't keep an eye on them, along with our mobile phones and anything else we had lying around. I had a vision of hordes of heavy-footed, light-fingered tradesmen invading our peaceful space, and felt a shiver of apprehension. Still, if not Omar then who?

But when we made him an offer he would only take a job rate, saying he wanted to be free to do other work. Just what we didn't want, someone who showed up for a day and then disappeared for five.

I had an awkward loose end to tie up: telling Hamza we wouldn't be engaging him to manage the restoration. I had been dreading this, not least because Hamza and his carpenter hadn't yet returned to assess the work we were still owed. If Hamza was no longer involved, there wouldn't be much incentive to ensure the work was finished.

But I needn't have worried. Hamza was gracious and affable. I started by saying that I knew he was terribly busy. Yes, he conceded, he was indeed busy. And, I continued, we knew his work was wonderful but his price was out of our league. We simply couldn't afford him. I saw him relax when I said this. I think he was secretly relieved that he didn't have to take on yet another job.

He was probably also thankful not to have me for a client. I'm the hands-on type who wants to know everything, that building managers dread. Most would prefer you to give them the money and then bugger off until the job is finished, with maybe an occasional pretend consultation before they go right ahead and do what they were going to anyway.

My meeting with Hamza ended amicably, our friendship intact, and I left feeling much relieved. He promised to come to the riad with his carpenter in the next few weeks to sort out the rest of the work.

ife in the Medina had its surprises. One evening around nine, there was a knock on the back door. As Sandy was out I ignored it, but then came insistent raps on the front door.

'Madame Suzanna,' a voice called out, echoed by that of a child.

I suddenly realised who it was and ran to let them in. Khadija was full of emotion, hugging me and showering kisses, and Ayoub turned his face up to be kissed. Our conversation was as complicated as ever, with her French worse than mine, but eventually her story came out.

Khadija's family had not moved to the countryside after all, but to a suburb on the outskirts of Fez. Her husband had lost his job and they had to move somewhere cheaper. As Abdul had seemed permanently stoned and was pretty haphazard when it came to work, I could well imagine that his employer had

had enough of him. Khadija was still embroidering slippers at home for a pittance to support the family. She asked if I had any cleaning work for her.

Having decided against employing Damia again, it was a timely request. And if things worked out I could probably find Khadija more work with other expats. I gave her the sheets I'd brought from Australia and she was thrilled. I think it was probably the first set of sheets she'd owned in her life. Ayoub was ecstatic about his paint set and immediately started using it. He would be starting school next year, Khadija told me, and she couldn't wait to have some free time to herself.

Their new place was an even tinier room in another shared house, and I gathered her son was driving her to distraction, and making it hard for her to do her embroidery. I thought, but didn't say, that having Abdul at home all day couldn't be helping.

Khadija came to clean a few days later, and all was well for the first hour. Then her dopey husband turned up with Ayoub in tow. He sat at the table chain-smoking and chattering incessantly, preventing her from doing her job properly. I didn't understand why, if he was out of work, he couldn't mind Ayoub somewhere else and let Khadija get on with it.

Abdul's presence also hindered my own work, a spot of editing for Sandy, who was on a tight publishing deadline with a novel. When Australian friends dropped in for a cup of coffee Abdul was taking up one of our scarce chairs. He eventually had the good grace to move, but sat two metres away on the edge of the fountain, staring intently while we talked.

After Khadija, Abdul and Ayoub had left, I noticed that a bright yellow hand towel was missing from the kitchen. It had been hanging in the centre of the wall at eye level, and the three of them were the only people besides us to have been in the kitchen that morning. It was such a stupid thing to steal that I wondered whether Khadija thought I hadn't paid her enough. Yet I had given her as much for three hours work as Abdul used to get for eight hours as a parking attendant. Sandy suggested she'd taken it home to wash, but it never reappeared. I hoped Abdul had been the one to take it, but there was no way of knowing for sure. I didn't care about the towel itself; it was the betrayal of trust that disappointed me.

Khadija turned up a few more times. I gave her tea but I didn't offer her any more work. She asked if we would employ Abdul on our restoration, but he hadn't a hope in hell. Eventually she stopped coming, and the last time I saw her was five months later, when she was pregnant once more.

While things with Khadija didn't work out, my friendship with Ayisha was flourishing, and we took to going to the *hammam* together. To get to her favourite one we had to walk through an area of the Medina called Ras Jnaan, which meant 'top of the garden'. Hundreds of years ago, it was the site of the market gardens that supplied the city. As Fez expanded, wealthier people moved from the city centre to build grand houses at Ras Jnaan, displacing the gardens. Now the area was forlorn and neglected, its houses in varying degrees of ruin.

Walking uphill, we clambered over a pile of rubble from a

recently collapsed dwelling that was blocking the alley. Half of the house still stood, a surreal cross-section with stairs that led nowhere and floors abruptly cut in two.

Further along, Ayisha pointed to a large dar on a corner where a friend of hers had lived when they were children. The family had moved away now and the house was uninhabited, its massive carved doors padlocked. I longed to go inside and take a look, and wondered whether someone would rescue it before it fell into terminal decline.

The house was so big and its corridors so dark, Ayisha confided to me, that she had been too terrified to stay overnight.

'I imagined the djinns were waiting for me,' she said.

'Tell me about these djinns,' I prompted. 'I've heard they live underground?'

'Oh, come on,' she laughed. 'I don't really believe in all that stuff.'

'I'm not asking if it's true or not. Just what people around here believe.'

She narrowed her eyes at me, as if assessing whether to go ahead, then nodded slowly. 'Well, some people think that if you go to a ruined building at night you'll see the djinns, and they'll hurt you. There are some houses nobody wants to live in because they're haunted. If you go there djinns can enter your spirit and possess you, and if that happens you don't sleep well. If you're still single you cannot get married. And if you are married you won't have children.'

I was intrigued. 'So what are they like, djinns? Are they the same as what we call ghosts, the spirits of dead people?'

She looked doubtful. 'Djinns are a kind of spirit, but they are independent beings, made by Allah from fire that does not smoke. Some people say they live under the ground, in communities. They are both male and female, with names and families, and they live much longer than us. They have their own stories. And they can be in conflict with each other, just like us. Djinns share everything with you, but they do things in the opposite way. They sleep during the day and come out at night. They find their way out into our world through places where there is water. That is why you must cover your drains.'

Moroccan squat toilets are plugged with stoppers, and drainage holes in floors and courtyards are covered with pieces of marble or large stones. I had always assumed that this is to prevent rats and cockroaches coming up from the sewers.

Ayisha's face showed how much she was enjoying this role reversal, the young woman imparting wisdom to the older. 'If you pour boiling water down drains or in the cracks in the floor you can hurt the djinns. Then you will be punished.'

Reaching the *hammam*, we stripped to our knickers, paid for and collected our buckets and went into the inner room to fill them.

'You cannot leave your children alone or a djinn might possess them,' Ayisha said thoughtfully as she scrubbed herself. 'Their faces will change, everything will change. They will grow up differently. I once saw a girl, only eight months old, crying in such an ugly voice. Nothing about her was normal. Her mother said she had left her for a few moments to go to the fountain in the street, and when she came back the child was in this state.'

'Couldn't the mother do anything?'

'She went to see a fakir, a Sufi who knows magic, and he told her to put her baby in a dirty, deserted place at night so the djinn would have a chance to return to its own kind. But the mother could not leave her daughter all alone for a night, and said she would accept her as she was. It was Allah's will.'

'If someone is mentally disturbed,' Ayisha went on, 'and talks to himself and does strange things, people think he is possessed by a djinn and can harm you.' She paused, unsure whether to continue, and when she did it was in a low, embarrassed whisper. 'Once, a few years ago, it was very hot and I was sleeping on the terrace with my mother. At two-thirty in the morning I opened my eyes and in front of me I saw a thin black man. As I watched he started to grow very, very tall. I was so afraid I could not speak, I could not cry out. I don't believe in djinns but I saw him. From that day on I could not sleep there.'

She covered her eyes at the memory and I could see she was disturbed by it. Time to change the subject.

'Are there other sorts of spirits besides djinns?'

'There are *marids* – they are the most powerful type of djinn. They are proud and arrogant. They are called blue djinns because their skin is blue and their hair always looks wet and wavy, as though they are under water. There are also *mlouks*, who can possess you simply because you are beautiful, or different in some way, or because they want what you have. So if you are different you better watch out.'

'And can people ever have friendships with djinns and those other spirits?'

'Yes. To make friends with djinns, you offer them milk or powdered henna, or burn incense. But you need to be careful because they have many tricks. If you need help you go to a fakir, and he will call on a *marid*. If I take a fakir a photo of the person I want to marry, he will ask a *marid* to fill this person's thoughts with me. Or if I have an enemy the *marid* can make trouble for them.'

Ayisha took a breath and followed this story with an even more extraordinary one. 'Sometimes,' she said, 'women will go to the cemetery and dig up the body of a newly buried person and cut off the hands. They use the hands to make couscous for someone they do not like. They give it to them as a present, and once that person eats the couscous he will never be happy again. He may even get sick and die. It is a curse.'

I'd never heard anything so macabre. 'But how do they use the dead hands to make couscous?'

'They take them home and wash them, then they hold them like so.' Ayisha reached across and took one of my hands in hers, stroking my palm with that of her other hand, as if rubbing butter through couscous.

I made a mental note to look twice at any unsolicited plates of couscous that appeared on our doorstep in future. Such a supernatural method of revenge cast neighbourhood disputes in a whole new light.

Ayisha's stories offered an insight into why women crowded around stalls in the souk that sold powders and potions, dried lizards, the skins of rare and endangered animals, live turtles and chameleons. They were seeking magical solutions to their problems from people they respected. Were they greatly dissimilar to those Westerners who bought endless self-help books, or tuned in to Dr Phil and Oprah for advice?

Just as Sandy and I were starting to feel completely despondent about our lack of a builder, fearing we were never going to get started, let alone finished, our luck turned. David rang to tell us about a wonderful builder who'd just become available; the project he was working on had stalled because of lack of money. Mustapha had worked on David's dar for six months the previous year, and was careful with sensitive preservation work such as ours. But he could be slow, David warned.

When Mustapha came to the door of Riad Zany I liked him immediately. He had a round, jolly face that made you want to smile, was dressed in a djellaba and skullcap, and was the epitome of gracefully aging Muslim respectability. He greeted me by gripping my hand in both of his. He was intelligent, assured and upbeat, and we felt confident about him, the more so since he'd been recommended by David, who was extremely particular.

Mustapha was willing to work on a day rate and we asked how much he wanted. It was a bit more than we'd budgeted for, but I reminded myself that we were talking about the equivalent of a few extra dollars a day. I said we needed to have the work finished in five months, to which he responded, 'That is a very short time,' a huge contrast to Omar's brash over-confidence. We happily engaged him to start at the end of May.

We had now hired the three people most important to the riad's future: Rachid, Mustapha, and the engineer, a woman called Zina. When we called a meeting a few days later Zina arrived wearing tight white pants and a green top with a black linen surcoat. She had streaked hair, heavy eyeliner, and high wedge shoes with diamantes. Needless to say, she also had a forthright manner, and she and Mustapha hit it off immediately. They had never met but had grown up in the same part of the Medina, and had animated conversations about what had happened to old so-and-so.

'I like this man,' Rachid declared, slapping Mustapha on the back. He and Zina had decided that Mustapha knew what he was talking about when it came to repairing houses, so they could relax.

We were standing on the terrace discussing how best to fix the catwalk, which had a rotting supporting beam, when some women leant over from the neighbouring house and asked Rachid something in Darija. The next moment all three were off racing downstairs and into the street like beagles after a scent. I followed at a more leisurely pace and discovered that one of the women had a house for sale and wanted them to see it. It was a nice little dar but unfortunately had been severely messed with — there was lots of badly applied paint and cheap tiles.

'This is just what I like,' said Rachid, tongue in cheek. I had to remind them all that they were doing a job for me, and reluctantly they dragged themselves away. We needed to see another neighbour to discuss repairing a shared wall, and had a great deal of difficulty locating their front door. Fassi houses are interwoven like some intricate puzzle, in a Lego method of construction. Our kitchen is partially over someone else's, and a neighbour's bedroom protrudes along one side of our courtyard, with one of our bedrooms on top of theirs in turn.

This sharing of space rarely involves being able to see into other people's homes, but it hints at the spiderweb of relationships that existed between the families who built and modified these houses over hundreds of years. Need a bit of quick cash? Flog off part of your house so the neighbour can build an extra room.

Our neighbour's door turned out to be in a completely different alley. We knocked and were answered by a woman who was understandably wary of a crowd of strangers, even if three of them were Moroccan, and she wasn't going to let us in for a moment. There was nothing wrong with her house, she said, she just wanted us to go away.

Mustapha asked to speak to her husband, but she told him he would only say the same thing. It was dispiriting, because we needed to see what was happening on the other side of our cracked wall. Since we were the ones paying to fix it, I'd have thought she'd be only too keen.

Zina had to leave for another appointment but Rachid stayed for a coffee, which gave me a chance to ask him why Fassi houses had developed the way they did.

'The Medina has grown without an urban plan,' he said. 'It followed agriculture and irrigation. Water finds its own way, and the Medina developed in a pattern like the veins in a leaf. The first roads were built next to the river because this was what people walked beside. They went there to get water for cooking or for their animals or to grow tomatoes. As areas further out were irrigated, people moved there, and as the canals and the streets running alongside them became smaller, they became increasingly private.'

I liked the image of the Medina as a leaf, with the central vein being the main route, and this leading into smaller ones, which were the *derbs*, or the private spaces.

'So why is the architecture here so different to the West?' I asked.

'Usually the shape of the parcel of land determines the architecture, but in the Medina they built houses from the inside, for symmetry, and what happened outside the building didn't matter at all. In the Western model, the inside is full and it is empty all around.'

I thought of suburban houses in the West with their vast expanses of lawn, essentially empty space, with the house as the central feature. My architect father had told me that when you design a house you should read the lie of the land, the way it slopes and the aspects of the sun, and then design the house to suit. It was a principle that wasn't always followed, and certainly not in the sprawling estates of poorly designed housing that were the bane of Western cities.

But why had this diametrically opposite approach developed in Arab countries?

'The Arabs were desert people and they found the emptiness frightening,' Rachid told me. 'So they created a way to formally control the space. The French philosopher Pascal said that when he saw the night sky and the stars it made him anxious, so he believed in God. Man needs something to measure by. In Fassi houses, symmetry is used as a way of making the space intelligible, based around the square or rectangle. *Zellij* is like a metaphysical science, used to make space measurable.'

I looked at the countless tiny tiles ranged across the courtyard in front of us, dividing the space into neat squares, five centimetres by five. Now they all made sense.

Things continued to go our way. Jenny rang one morning to say she was on her way over with our new plumber. She warned me not to shake his hand and I gathered he was a fundamentalist. At least his scruples didn't extend to not working for infidels. When they arrived he poked and prodded walls and had a go at dismantling the fountain, blowing through the nozzle to see if the channel was clear. We began to feel hopeful he might actually be able to fix it.

After the plumber had left, Jenny and I went to purchase building supplies, trailing some distance behind Si Mohamed. He was a neatly dressed, mosque-going fellow in his early twenties, tall and lanky with a ready smile and a helpful sincerity. His father had died when Si Mohamed was young and he was now the sole breadwinner for his mother and two sisters. Staying well behind him in the Medina was imperative because the previous year he had been arrested for trying to sell stuffed toy camels to foreigners and spent two months in prison.

PREVIOUS   A view over the Fez Medina, with the green-tiled
roof of the Karaouiyine Mosque in the foreground

ABOVE   The view from our terrace

RIGHT   The downstairs salon on the day we purchased the house

ABOVE One of Fez's last surviving
brocade makers

BELOW 'Liberace', at the café where
he's worked for forty years

RIGHT Perfume oils and sacred
and medicinal herbs for sale

FOLLOWING The courtyard, for
many months our living room,
was in constant chaos

ABOVE The decapo ladies, Fatima and Halima, whose work revealed the beauty of the cedar beneath up to seven layers of paint

BELOW The first of many deliveries of lime and sand that were to continue for months

ABOVE The kitchen stripped bare. The outline of the old window shows where it once joined the house next door.

BELOW Putting zellij on the kitchen floor

ABOVE   After months of work, the courtyard is finally complete

LEFT TOP   The difficult and dangerous task of lifting the catwalk
             balustrade into place

LEFT BELOW   Looking down into the completed courtyard

FOLLOWING   The finished downstairs salon. The door at the end leads to
             the main bathroom, the door on the right to the courtyard

ABOVE The view looking up from the kitchen through the completed halka

BELOW The kitchen, finished at last.

Over the past couple of years, thousands of false guides had been imprisoned for months at a time. They were mostly young men who pursued tourists and wouldn't take no for an answer, and they could ruin a pleasant stroll through the Medina. Frazzled tourists often paid a few dirhams to get rid of them, which only exacerbated the problem. Although imprisonment was an extreme solution, it had the desired effect, and now there were fewer prepared to risk it. The downside was that any Moroccan seen walking with a foreigner could be arrested, which made life difficult if you had Moroccan friends or employees.

The building-supply merchants were in Bab Guissa, a very old area of the Medina I had never visited before. Its narrow streets teemed with people. We went past a number of hole-in-the-wall shops until we came to a place the size of a garage, filled with bags of sand, lime, cement, and a pair of old-fashioned scales. To open an account there, I just needed to write down my address. No ID was required, no proof of address or passport. About ten or so donkeys were standing patiently by, ready to make deliveries. They would bring our initial order of lime and sand the following day.

We went on to a wood supplier, where long lengths of cedar reached up to a soaring ceiling, making the air smell like a forest. I bought four large sheets of masonite to protect the tiles in the courtyard. These were neatly rolled up and a small, wiry porter of about sixty lifted them deftly onto his head. Porters are often employed to carry awkwardly shaped objects, or those not considered large enough for a donkey.

Jenny and I had to run to keep up with him as he raced

off downhill and through the tiny streets, cutting through the crowds like an escaped bull. We lost sight of him after a few twists and turns, but he was waiting at the riad's front door when we arrived ten minutes later. These porters prided themselves on knowing every one of the Medina's thousands of alleys.

At last everything was in place and we were ready to begin. Saturday, May 27 was a red-letter day – the first day of construction. Early that morning, I went to fetch some bread and found three men sitting on a step in our neighbourhood square.

'*Salaam Aleikum,*' I said, wondering what they were waiting for.

'*Aleikum Salaam,*' came the chorused reply.

They were still there when I returned, and moments after I got home there was a knock on the door. I opened it to find the same three men there – our builder and his two labourers. I hadn't recognised Mustapha in his work clothes.

The three of them proceeded to empty the kitchen. Within minutes our stove, crockery and utensils were sitting in the courtyard. Then they began to rip the plaster off the walls. The speed of their work, along with the noise and the clouds of dust, had us dumbstruck. After so many weeks of frustrating delay, it all seemed to be happening at once, and Sandy and I didn't know what had hit us. We quietly regretted not having had our coffee half an hour earlier.

By lunchtime, most of the plaster was off the walls, and the skeleton of the kitchen had begun to reveal itself. It was more than just intriguing ancient brickwork – there were outlines of several windows that had been filled in a long time ago, including one through to the neighbour's house.

An insistent braying from the alleyway announced the arrival of a team of donkeys, making their first delivery of lime and sand. As one donkey was being unloaded it decided to make a break for it, clattering down the steps of the alley, panniers flapping. I would have felt like doing much the same thing under the circumstances. It was brought back and given a reprimand with a stick by its exasperated driver, who no doubt had a schedule to adhere to.

After lunch, Mustapha and his team lined a corner of the courtyard with masonite and then covered it with plastic. Onto this they shovelled huge quantities of lime and sand, piling it into the shape of a volcano. Then they ran a hose from the bathroom and started filling up the cone. This was the beginning of the traditional lime-and-sand mortar undercoat known as *haarsh*, which is used on interior walls.

Fassi building methods evolved to suit the natural conditions, but they take time, something many people do not want to spend these days. *Haarsh* used to be allowed to cure for three months or so. You made your *haarsh* and then went off on your *Hajj* – pilgrimage to Mecca – on your camel. By the time you returned, the mix was cured. Nowadays builders, even traditional ones, only cure it for a few days, before adding a small quantity of cement to get it to set.

But cement does not allow the walls to breathe in the same way that traditional plaster does. Moreover as cement contains lime, you risk creating an excessively lime-rich mixture, resulting in walls that will crumble in a few years – a bit like the 'concrete cancer' problems in many Western buildings more than twenty years old. What is the amount of cement that can be added before this happens? No one knows.

⌣

Sandy and I quickly recovered from our initial shock and were thrilled that things were happening at last. It would be months before the house returned to anything like its former glory, with a few modern adaptations like a decent kitchen, but we were on our way.

By the second day, the kitchen looked like a different room. And amazingly, the two major structural problems that we had thought would take so much time and money had evaporated. The bowed wall and the collapsing ceiling were in fact illusions. A false wall had been constructed out of cardboard to hide the top half of a storage mezzanine. And when Mustapha and his men took the plaster off the ceiling, magnificent cedar beams were revealed, completely intact. They formed a central square, the shape of a *halka* – a traditional open hole through to the room above, which at some stage had been covered in.

There was about a foot of rubble on the kitchen floor, but the room now felt twice as big. Behind the cardboard wall, we discovered not just more space but a collection of treasure and trash. We were thrilled to find a tall earthenware urn, intact and complete with sticky brown remnants of the olive oil it once stored. It looked like a Greek amphora without the handles. Then we pulled out the remains of an old crystal-set radio, a fishing rod, a rusted metal bucket with a lid, a soccer ball, an old picture of the King, and a woven basket from southern Morocco, used for collecting dates. There were also piles of rotten wood, which we planned on donating to the local bakery's kilns.

As the demolition progressed over the next few days, it became clear that we were employing too many chiefs and not enough Indians. Apart from Sandy and me, in the way of overseers there were Jon, Jenny, Rachid, Zina and Si Mohamed. Only three people were actually doing the physical labour, but now that the management structure was set up it was difficult to change. On the plus side, Mustapha, whose real work would begin when the reconstruction started, spent time teaching Sandy Darija. In the meantime his two workers, Abdul Ramin and Mernissi, were the ones doing the banging, digging, hefting and carrying.

Mustapha had asked me to buy gunpowder tea, the sort that is mixed with fresh mint to create the tea Moroccans drink in copious quantities. I did this, and he seemed to have the idea that I should make it. I resisted, not wanting to get trapped in the role of tea lady to the workers. As Si Mohamed was sitting around watching the work, I palmed the job off on him.

'Moroccan whiskey,' said Mustapha, sipping the over-sweet brew. He claimed that with this tea they didn't need to eat. They worked through lunch and left an hour early, at four o'clock.

Sandy had just finished washing down the steps to the front entrance, in an attempt to stop sand and lime coming inside, when another donkey train arrived with yet more sand. As the workers had left for the day, we got the drivers to deposit it outside the door. Several young boys in the alley were taking a great interest in proceedings, so Sandy asked them if they wanted to clean the street of the spilt sand for a few dirhams. They were keen, and he gave them a hose. Big mistake. They had a lot of fun, generating a great deal of muddy water, which ran down the hill and under the front door of a house further down the alley.

I had often seen its owner, a tiny woman, standing on her doorstep and staring along the street with vacant eyes, as though waiting for someone or something that never arrived. She ignored me whenever I greeted her, but now she popped out of her door like a jack-in-the-box, yelling and screaming at the kids, who all scarpered, leaving Sandy standing there helpless. He did not have the words to apologise to her or say he would fix the problem.

Hearing the commotion, another neighbour put his head out, and after assessing the situation, came and rescued Sandy by taking a broom and marching down to her doorstep. Sandy fetched another broom and joined him. They cleaned her steps and front entrance, the woman muttering at them all the while. When they'd finished and she'd closed her door, the neighbour pointed to his head and tapped, indicating she was crazy.

'*Fou!*' he said, just in case Sandy hadn't understood.

Sick of living out of a suitcase, I resolved to find a wardrobe to keep our clothes out of the dust. My starting point was the workshop of a local carpenter, an Aladdin's cave of old doors taken from Fassi houses. Those houses must have been very grand, for the doors were carved cedar, over three and a half metres in height. Their original owners had either decided they needed the money or wanted to modernise. And of course there was also the possibility that some had been stolen.

The carpenter was a young man who looked like Cat Stevens in his fundamentalist phase, with a black beard, white surcoat and skullcap.

'Who will buy all these doors?' I asked him.

'God willing, we leave for Marrakesh this night,' he said, making it sound like he was off on a voyage of many months. I understood immediately: the doors were being sold to wealthy foreigners. I suspected that many of them would end up in New York or Paris apartments – more pieces of cultural heritage lost to Fez.

Among all the gigantic doors, one set stood out. They were tiny. Made of heavy cedar, they were so small they looked like they'd been made for a race of hobbit-sized folk. But more than that, they were a technicolour blast. Layer after layer of paint had been scraped back in a haphazard fashion to reveal patches of orange, blue, yellow and various shades of green. They had the exuberance of an abstract painting, all the more interesting because the art was unintentional – the result of decades of utilitarian renewal.

Quick to spot my interest, the carpenter was at my shoulder. 'Those come from the Mellah,' he said.

Sandy and I had walked through the Mellah, the old Jewish quarter in New Fez, a few days earlier. Once a thriving community had lived there, but now only a couple of hundred Jewish people remained in Fez, and most had relocated to the Ville Nouvelle. Some houses in the Mellah still looked lived in and cared for, but many were in a sad state of disrepair.

One house in particular had caught our eye. It was of graceful proportions with stylishly carved window frames and balconies. Jewish houses in Fez are unusual in having their balconies on the outside, there being no need for women to be shielded from passers-by. No longer occupied, this house was sliding into an irrevocable decline. Through a shutter hanging drunkenly off a window, we glimpsed a room with beautiful plasterwork, similar to that in our *massreiya*, now exposed to the elements.

'It's a pity there's not some sort of heritage organisation to restore houses like this,' Sandy mused. 'If they've been empty for, say, fifty years, they could be rented out to pay for the work. Any surplus could go back into the community.'

It wasn't a bad idea. 'But what if the descendants of the owners returned after a few years?'

'It could be held in trust for them,' Sandy said. 'If the community has an interest in maintaining the house, then it will be conserved.'

Jews had begun to settle in Fez from its earliest days, and their fortunes waxed and waned depending on the attitudes of those in power. In the ninth century, a raft of restrictions were imposed on them, including a decree that they could only wear black clothes and sandals, no shoes. They could not ride horses and had to pay higher taxes than everybody else. When Fez was conquered by extremists in 1035, six thousand of their number were massacred.

But by the thirteenth century, when the Berber Merind dynasty ruled, Jews were protected and their businesses thrived.

The Mellah was established in 1438, just outside Fez, ostensibly to shield Jews from the Muslim populace. The site of the Mellah had previously been known as al-Mallah, meaning 'salty area', and eventually the word 'Mellah' came to refer to Jewish quarters in other Moroccan cities as well. The explanation for the word's origins has changed over time, and today you might be told that the Jews, in return for their protection, salted the decapitated heads of those who rebelled against the sultan, in preparation for a grisly display on the city's walls. Although the practice of displaying heads on the walls survived into the twentieth century, I couldn't find any evidence for the salting story.

Tolerance levels continued to fluctuate, particularly when the Merind dynasty fell, and then again at the end of the eighteenth century when the entire Jewish community was expelled from Fez and the synagogues destroyed. Two years later they were permitted to come back, but only a quarter returned.

The French ruling classes of the early twentieth century were hardly sympathetic to the Jews, as the notorious Dreyfus Affair demonstrated, and when the protectorate began in 1912, violence flared in the Mellah. Although Jews were not deported from Morocco during World War II, they suffered humiliation under the Vichy government, and many chose to emigrate when the state of Israel was created.

Nowadays, Morocco has a reputation for being the most tolerant of all the Islamic countries towards its Jewish population, which is around five thousand nationally, and many Moroccans will tell you proudly that the two groups live in harmony. Morocco is the only country that allows Jews to retain dual citizenship after emigrating to Israel, and many still return each year for religious festivals. Several have held high positions in business and government.

I was captivated by the doors from the Mellah in the carpentry workshop. I could see years of history in them, a multitude of stories of daily lives through the generations. But, I reminded myself, I was here to look for a wardrobe. The last thing I needed was a set of doors I had no use for.

In a corner of the shop, I spotted an old cupboard with carved front panels being used as storage for the carpenter's tools. On closer inspection, I saw it had been damaged and badly patched in numerous places. It would do. When I asked the price I was told an astronomical amount.

'You will see something like this in the museum,' the carpenter murmured near my ear.

Maybe, but at that price I would be paying more for the wood-worm and holes than for the actual wood. I bargained him down half-heartedly, then said I was going to fetch my husband. I was unsure about the cupboard but keen to get a photo of the doors.

When I returned with Sandy in tow he was immediately struck by the doors, in the same way I had been. I took a couple of photos of them and then showed him the cupboard. He was unenthusiastic and turned back to the doors, asking how much they were. The carpenter told him a price that would have paid for the refit of our kitchen. We smiled and shook our heads, saying we couldn't afford them, and walked out.

Sandy was already at the bottom of the hill when I made the mistake of pausing and looking back.

'What are you thinking?' the carpenter called, sensing a sign of weakness. 'Is it not a pity that you do not have such doors?'

'Yes,' I agreed. 'It is.'

'You say how much you will pay.'

'Two thousand, five hundred dirhams,' I said. I didn't have to look over my shoulder to know that Sandy had heard this and was shaking his head in disbelief at my audacity, as this was a fraction of the price quoted.

'Seven thousand,' said the carpenter.

'Two thousand, six hundred.'

'Let us be serious,' he said. 'Six thousand.'

This went on until the carpenter, seeing I wasn't going to shift much further, accepted three and a half thousand dirhams and the doors were ours. Although pleased with the deal, I had mixed feelings, wondering if by buying the doors I was fuelling a market in cultural heritage that I did not agree with. I consoled myself with the thought that at least these two remarkable doors would remain in Fez.

'But what on earth are we going to do with them?' Sandy said as we left the shop after paying for them and arranging delivery.

I had no idea. We now had a useless set of beautiful doors but still no wardrobe for our clothes, which along with everything else in the house were coated in a fine film of lime and sand. When I woke coughing in the middle of the night I thought about the damage microscopic particles can do to the delicate tissue of the lungs. We had brought expensive dust masks back from Australia for the workers but they refused to wear them, saying they'd be too hot. We cajoled and threatened, painting a picture of what would happen to their health in the future if they didn't use them. But we couldn't get them to take us seriously.

One day, I was horrified to see the men breaking up an asbestos drainpipe in the courtyard with a hammer – dust and fragments were flying everywhere. They were disbelieving when I told them what harm this could do. More as a reaction to my semi-hysteria than from any understanding of the implications, they stopped smashing up the pipe. But they still would not don masks for the dusty jobs.

I quizzed Si Mohamed about this. 'Why won't the men wear masks?'

He shrugged. 'They are Moroccan men.'

What was he saying? That Westerners were pampered pansies for trying to avoid mesothelioma? There wasn't a culture of workplace health and safety in Morocco. Despite numerous regulations, there were insufficient resources to implement them. More than seventy per cent of workers did not have medical insurance. We chose to pay if our workers needed to visit a doctor or dentist, knowing they could not afford to.

On the other hand, Mustapha told me about one of his builder friends who, aged 105, had married a woman of 101 the previous week. It was good to hear that there were some long-lived builders around, although that probably gave the rest of them a false sense of security.

Being concerned for our own health as well, we had intended to move out of the house for the most intense part of the restoration, but the Fez Festival of World Sacred Music was about to begin and all the accommodation was full. The only inhabitant of the house who didn't seem fazed by the environmental hazards was the kitten we'd acquired.

I had resolved not to rescue any of the pitiful, mewing little balls of fluff I'd seen on the streets, steeling myself to walk past them even though the riad was cat paradise – a courtyard with trees, a house with lots of nooks and crannies, and a regular food supply. But we had to go back to Australia, and I wasn't about to adopt something I couldn't promise a future to.

Then one evening we were invited to a barbeque at the home of Peter and Karen, an Australian couple in the Medina. They had a house up the hill from us, a tall and narrow dar with formal rooms downstairs and the main living area on the first floor, surrounding the four sides of an atrium. It was when we ascended to the roof that we discovered their reason for buying the house – the view was stunning. It was like being perched in the centre of an amphitheatre, with the panorama encompassing the west of the city. Alpine swifts dived and circled in the darkening light. Over the terrace wall, we could see the goings on in the local square.

There was a sense of the surreal having a barbeque in such a location. While we were eating, a tortoiseshell kitten came out of her hidey hole on the terrace and started to play with the straps of Sandy's bag. She had bald patches on her head, making her look as if she'd had brain surgery.

A few weeks before, Karen had seen her sitting on the street looking pathetic and miserable. She was covered in globs of glue, and although Karen too had resolved not to rescue kittens, she was unable to resist taking the little thing home to clean her up. The only way she could get rid of the glue was to pull it off, hence the bald patches. The kitten had flourished with Karen and Peter's care and was playful and trusting, but they were returning to Australia the following week and dreaded having to put her back on the street.

Being suckers for small creatures, Sandy and I offered to look after her for the five months until their next stay. When she climbed up my leg and settled on my lap, falling fast asleep, I decided she'd be the perfect Fez pet – a returnable one.

Our chaotic new life began to develop a pattern. Most mornings found Si Mohamed and me running back and forth between home and Bab Guissa, shopping for supplies for Mustapha, while Sandy stayed home to supervise. One day, Mustapha gave me an order for five hundred bricks, but I discovered to my amazement that there were no first-quality, handmade bricks to be found in Fez. A woman in Marrakesh had bought them all, and as only one factory made such bricks, it would take a few days to restock.

On our way back to the riad, we passed a house that had just collapsed. Emergency workers were clearing the rubble while a curious crowd packed the street, hampering their efforts. Si Mohamed asked a bystander what had happened, and we learnt that a pregnant woman had been killed when the roof collapsed. Such occurrences were unfortunately regular in Fez, although after the collapse which killed the eleven mosque worshippers, the authorities had erected scaffolding to prop up more than a thousand endangered buildings.

I returned home with a renewed respect for the dangers of cracked walls and rotten beams, and found two potential new employees waiting for an interview. Fatima and Halima were, as David put it when he recommended them, 'lady strippers'. They stripped paint off timber and ironwork with Decapant, a chemical paint remover. Both in their forties, they were dressed, like most women of their generation, in traditional style, but had been forced into a non-traditional role because their husbands were sick and they had children to support.

Fatima, the more assertive of the two, had a narrow face and a beaky nose. Halima's face was softer and round, with two teeth crossed over at the front that gave her a slightly goofy look. But appearances were hardly relevant. They were keen, and the idea of employing women appealed to us. It was rare to find women working in the building trade in Morocco – it was unusual anywhere, for that matter.

I told them we'd give them a week's trial, and after that we'd see. As few Moroccan workers bring so much as a screwdriver to work, it was no surprise to find these confident women giving Si Mohamed a list of things they needed. I followed him to our local hole-in-the-wall hardware store at R'Cif to pay for it all. It was just as well I did, for when the list was being assembled I noticed two huge wire brushes on the counter. I stared at them in alarm. Were these what the women intended to use on our delicately carved wood? It would be shredded in seconds. When I questioned Si Mohamed he argued with me.

'These are what everyone uses for Decapant.'

Not convinced, I asked him to wait while I called David, who couldn't have been more emphatic.

'Do not let a wire brush into your house,' he said. 'Absolutely not. Get soft steel wool instead. Just ask for *halfa*.'

Armed with my newest Darija word, I completed our purchases, but when Fatima and Halima started the next day, they were disgusted I wouldn't let them use wire brushes. It was the beginning of a tussle that even saw them smuggling in the banned implements. But it was a tussle I won.

I was tired of having to trail through the Medina twenty paces behind Si Mohamed so he wouldn't get arrested again, but there was no official permit that would allow him to be seen publicly with us. So we devised a strategy for creating one.

I wrote a letter stating that Si Mohamed worked for Sandy and me and needed to be on the street in our company when we went to buy materials. He translated it into Arabic, and to the relief of us both it was duly stamped and authorised at the government office.

'Do we need to do more than this?' I asked the man behind the counter, and was told no, that would be fine.

Walking out of the office, Si Mohamed and I headed for Bab Fettouh to buy sponge to make mattresses for banquettes. As we got out of the taxi, two plainclothes police accosted us, demanding to know what Si Mohamed was doing with a foreigner. He pulled out the freshly stamped authorisation and they scrutinised it, trying to find something wrong. Reluctantly they handed it back and told us we needed to go to the main police station to register it.

Deciding we should play by the rules, no matter how absurd, we caught another taxi to the police station in the Ville Nouvelle. We were shunted from office to office before we found a man who said that if I were working for Si Mohamed he could help us, but as it was the other way around we needed to go to the Tourist Police.

We went back to the centre of town in yet another taxi and into another police station. We knocked on several doors but it seemed no one was in. We went for morning coffee, and on our return found the office of the Chief of Tourist Police, explained what we wanted to his secretary, then waited outside while a family went in. Finally the secretary poked his head out and made a gesture to come in, but when I moved forward he put up a hand to stop me. Si Mohamed went in alone.

Moments later he was back out, shaking with impotent rage. 'It is useless,' he said.

As we rode despondently home he told me the story. The Chief of Tourist Police turned out to be the man who'd put him in jail the previous year and he remembered him. Si Mohamed explained his new situation but the chief shouted him down, insisting that if he was seen in the Medina with a foreigner he'd be arrested.

When Si Mohamed asked the chief what I should do when I needed translation help he was told I should hire a licensed guide.

'But Suzanna knows her way around the Medina as well as I do,' Si Mohamed said.

'She still needs a guide.'

I'd never heard anything so ridiculous. I was supposed to ring a guide every time I wanted to pop out for a few building materials? I didn't need someone to show me where to go, just someone who understood Moroccan building terms and what it was our workers needed. Besides, what guide was going to be on call for half an hour's work at a time, several times a day, at reasonable rates?

I asked around to see what other foreigners did and was told that all the workers were supposed to stay inside the house. If they were seen on the street with you they'd be arrested. I knew the crackdown on guides had been good for tourism, but this seemed a tad overzealous. What if Ayisha were seen on the street with me? I wondered.

As it happened, I'd been out with Ayisha a day or so before. We stopped at her house to drop off her shopping, since it was close to our riad. I suppressed a gasp of surprise when I entered her house – I had not been in a smaller, poorer house in the Medina. The downstairs courtyard was minute, lined with uneven concrete, and Ayisha's family rented a room upstairs, a space of about twenty square metres. This was the entire space for a family of seven. At one end was a niche with an impossibly tiny kitchen.

'Where do you all sleep?' I asked.

Ayisha waved at the banquettes around the room. 'Five of us sleep here, and the other two up there.' She indicated a small, curtained mezzanine. The space was hardly big enough for seven people to stretch out full-length, let alone have a modicum of privacy. Such poverty did not equate with the beautiful, well-dressed Ayisha I knew, yet it was the only house she had ever lived in. When she'd told me she came from a poor family I had not realised how poor.

On the way out, I was confronted with two entrances. I went to take the one that led to the street, but Ayisha stopped me.

'No, not that way,' she said. 'That way is only for dead people.'

'What?' It was an ordinary-looking door leading straight onto the street from the courtyard, whereas the main front door, four metres along, was screened by wooden panels so that passers-by could not see inside.

'They only use that door when someone dies and they need to get the body out.' Ayisha made her body taut, holding her hands up to her chest. 'This is the way they take you out, because of our religion.'

She told me that when someone died their body was put on view for the family, after first being ritually prepared by those of the same gender. It was washed with perfumed oils, the feet were bound, and the right hand was placed over the left on the chest.

'They wrap the body in a piece of white cloth and they put it in the middle of the courtyard and read the Koran on his soul,' explained Ayisha. 'Then they put it on a plank of wood and take it out the door to the cemetery. Women are not allowed to go with it, they can only visit the grave on the second day.'

I was grateful for the way Ayisha would volunteer these insights into Moroccan customs. Shortly afterwards I was the one giving her the insights – on the mysteries of academic writing when she asked me to correct an English assignment for her. She had almost completed her arts degree and was understandably nervous about her job prospects once she graduated. Morocco has tens of thousands of unemployed graduates competing for an extremely limited pool of positions, but that didn't deter students from working hard to obtain a degree – on the contrary.

On our way home through the Medina one night, Sandy and I were bemused to see groups of young men gathered under streetlights, reading. What were they doing? The mystery was solved by David the next day. It was close to exam time and people studied on the street because they had no private space to do so at home. Nor could they turn on a light when everyone else was sleeping. I wished that a number of Australian students I knew could see them.

I suggested Ayisha write a CV, backing it up with some references, which I could give to the owners of guesthouses I knew. I thought she'd make an ideal front of-house-person. When I told her she was welcome to come to a tutorial Jon was giving me on Excel spreadsheet, so she could include it on her CV, she inexplicably burst into tears, putting her head on my shoulder and sobbing.

I comforted her for a moment and then went on in a matter-of-fact way. With my Anglo-Celtic reserve, I didn't know quite what to do with the weepy young Ayisha, who, like many Moroccan women I'd met, was far more used to openly expressing her emotions than I was.

At the end of their first week of work, I sat down at a table to pay the workers in the customary way. Si Mohamed called them individually and explained the amounts to them.

'Six days at a hundred and fifty dirhams a day is nine hundred dirhams.'

I handed them the money and waited while they counted it, then thanked them for their work.

'You look the picture of a colonial administrator,' Sandy laughed, and for a brief uncomfortable moment, I had a vision of myself as the patron of a rubber plantation, doling out a pittance to exploited workers. Yet the reality was that we were giving six people full-time work for several months on above-average wages. Local unemployment was so high you could virtually walk out into the street and say you wanted unskilled labourers and you'd have an instant queue. It was finding people with traditional building skills that was difficult.

We liked our workers. All were over the age of forty, except Si Mohamed, and they had a steady, reliable air. The men were usually silent while they worked, the only sound a rhythmic *chink, chink* as they removed plaster. Fatima and Halima, whom we'd dubbed the decapo ladies, maintained a constant chatter, and as I couldn't understand what they were saying, it was a pleasant background noise that could almost lull me to sleep. There was a wonderful sense of purposefulness in the house, of a shared goal.

'Your house is my house,' Mustapha declared one day. 'I am working from the heart.'

The riad had not had any proper maintenance done for decades. Now was its time.

The following week, the plumber came to repair the fountain. We watched with concern as he began to dig a channel through our beautifully tiled courtyard to find the pipe.

I had ignored David's disparaging comments about the Victorian style of the fountain. 'It's not at all traditional,' he protested. 'Fassi fountains were very low.'

Sandy was all set to take it out, but I wanted to wait until we had something better to replace it with. Antique marble fountains being rare and expensive, I knew we were in for a long wait.

But even the prospect of sitting in the courtyard with the sparrows chirping to the sound of tinkling water did little to assuage our dismay at the destruction of sections of the courtyard. Thankfully the plumber came across the pipe after digging only a couple of small holes. It was completely rusted up. He removed and replaced it.

He made a couple more visits, installed a small pump, and when he switched it on everyone stood around to watch, clapping and cheering as the water shot skywards. Sandy adjusted the pressure to a gentle spray, which fell into the bowl and splashed over the edge into the pool below. It was so pretty that we spent hours at night watching the play of water – much more restful than television.

Meanwhile more discoveries were being made. Beneath the dull grey paint on the *massreiya* doors, the decapo ladies uncovered an exquisite geometric design painted in yellows and browns. We estimated it dated from the mid-nineteenth century, about the same time the Iraqi glass windows in that room were created. The women had already damaged one section by leaving paint stripper on it overnight, so we halted work on it until we got some advice.

David came to the rescue yet again and recommended a restorer. Using cotton buds, the man painstakingly removed layer after layer of paint. It took three days before the design was revealed, and fortunately it was largely intact. It amazed me that someone had painted over it in the first place.

The Fez Festival of World Sacred Music, held during June, was a reminder of what the world outside had to offer. It had begun in response to the first Gulf War of 1991, as a way of bringing different religions together to share and appreciate one another's traditions. This year's nine-day festival had a smorgasbord of spiritual and religious music from Syria, Iran, India, Mali, Latin America, Japan, Tibet, Azerbaijan and the Mediterranean, along with a talkfest on such topics as wealth and poverty, spirituality and ecology; a literary café; and documentaries that were to be presented by the filmmakers.

As the writer and photojournalist behind 'The View from Fez' weblog, Sandy and I were given media passes in order to do reviews; I was also working on a travel story for my paper in Australia. Fortunately for us, the concerts began in the late afternoon, after our workers had left for the day. They were held beneath the spreading branches of a giant oak tree in the courtyard of the old Batha Palace, which was now a museum of Moroccan arts. The evening concerts were held at Bab Makina, an enormous space with crenellated walls at one of the entrances to the palace, about ten minutes' walk from Bab Bou Jeloud.

Built in 1308, Bab Makina had spent part of its life as an armaments factory, so it seemed appropriate that it should be the main venue for an event promoting cross-cultural understanding. The audiences were a mix of upper-class Moroccans and Europeans, mainly French. The high prices for the tickets subsidised free concerts for poorer people in a square near Bab Bou Jeloud.

On the first afternoon, I watched in fascination as a whirling dervish from Syria began to rotate slowly in the palace courtyard. There wasn't a lot of room and I wondered if he would spin off like a top into the audience. His eyes closed, a peaceful expression on his face, he spun faster and faster, arms akimbo, until the hem of his long white tunic lifted to form a complete circle. As it wasn't possible to see his feet, it seemed as if he were levitating, about to take off to another realm entirely.

Late every evening, a Sufi group performed. Sufism, the mystical branch of Islam, is focused on letting go of all notions of the individual self in order to realise unity with the divine, and many Sufi groups use music and dance to achieve an ecstatic state. This face of Islam is very different from the hardline Wahhabism that has parts of the Middle East and Africa in its grip. Whereas Wahhabists must deal with God through intermediaries, Sufis feel they have a personal connection with God – a God who is loving, not a judgemental deity to be feared. Sufi brotherhoods are particularly prevalent in Fez.

The Sufi performances were extremely popular and getting in wasn't easy. The night we went, there was such a crush of people that Sandy opted to stay in the gardens, but I needed to get as close as possible to take photos. Despite having a media pass, I had to argue the point with a security man, who eventually took pity on me and led me to the front of the queue. I worked my way slowly through the crowd, trying not to step on toes or babies.

The group performing was a local Fassi brotherhood who regularly played at weddings and circumcisions, and the audience was mostly locals, families with kids, teenage boys with tight T-shirts, veiled women. I found a spot wedged beneath a huge fountain and a French camera crew, and ended up almost sitting in the sound guy's lap. The proximity of so many people was overwhelming. It was easy to see how people get crushed to death by the sheer force of numbers on pilgrimages to Mecca.

When it came time for the musicians to enter they were led by an angelic young boy carrying a candle, and people cleared a path to let them through. The musicians stood in a line facing the audience, and the chanting, drumming and shaking began. The oldest Sufi, a wrinkled prune of a man, had his eyes closed, lost in the rhythm. The singers were dressed in pale robes and sat on one side of the stage, facing a row of musicians in elaborate woven tunics with red and white stripes. A man in the middle led the refrain, keeping time with a small drum.

The songs were sung very loud, and to an untrained ear sounded a little tuneless, but the rhythm and repetition and sheer energy they exuded was exhilarating. The audience swayed as they sang along to what were obviously familiar numbers. Little girls tossed their hair around wildly, and even demurely dressed matrons looked as if they were about to faint. Every song pushed the outpouring of energy up a notch, until the noise and dancing of the crowd were at fever point.

Then two young musicians rose and blew with such ferocity on their long silver trumpets that those close by were in danger of suffering permanent hearing damage. In the brief lull that followed the trumpet blasts, I made my escape, to avoid becoming a human pretzel. People were still trying to squeeze into the already over-crowded area and meeting resistance from the security staff. One male journalist being denied access was on the verge of fisticuffs. I slipped past into the night just as the shouting reached crescendo level.

Afterwards, wondering what it all meant, I sought out the maker of a documentary called *Sufi Soul*, which was being shown at the festival. Simon Broughton, who also edits a world music magazine, told me there are some fifteen main Sufi brotherhoods in Morocco, each with its own *tariqa*, or 'way'. One of the most popular, and also the most flamboyant and theatrical, is the Ais-sawa brotherhood, the group we had seen perform.

Simon knew a member of the Aissawa brotherhood who worked at the Chouwara tanneries, and for him the hard work there was made bearable by the thought of performing after it was done. 'The music is intended to bring the djinns in, not drive them out,' Simon told me. 'They're brought in to deal with specific needs and problems – to relieve anxieties and tension, for example.

Different people need different djinns. Sometimes a djinn is summoned by name. The musicians might call on the lady Aisha, and a woman in the audience will jump up, seized by Aisha. It's believed that the djinn has manifested in her.'

I had heard that Aisha is supposed to be beautiful and seductive, but with the hooves of a goat. She appears to men in dreams, and sometimes when they're awake. Children are often afraid of her.

According to Simon, the majority of Moroccan Sufi bands are male, although there are plenty of female devotees. 'And there are all-female groups,' he said, 'some of which have been around a long while.'

He hadn't seen any mixed-gender bands, but he told me that in the Jilala sect women are heavily involved as singers and dancers alongside male musicians. The Jilala are usually seen at *moussems*, a local festival held to honour a saint or holy man. 'They get the women dancing,' said Simon, 'and frequently the women go into a trance.'

Such trances also occur when other brotherhoods perform. From what I saw at subsequent ceremonies, the state appears completely spontaneous, an altered consciousness of which the participant, usually a woman, has no recollection afterward.

While my first experience of a Sufi performance was intriguing, my concern about being crushed had made it difficult to enjoy. So I felt privileged when, some time later, Sandy and I were invited to a private Sufi healing ceremony for an English friend with cancer.

The ceremony was held in a riad in the backstreets of the Medina. We arrived just before ten at night to find carpets laid out in the courtyard, with sheepskins and cushions to sit on. This time there were just twenty of us in the audience, mainly Europeans. We took our places for the entrance of the Aissawa Sufis, who were wearing turbans and colourful tunics, some carrying maroon and green flags, others blasting their silver trumpets into the night. Arranging themselves in a circle, they sat down and began to play drums, *ghitas* (a smaller version of a trumpet), and clackers.

At first I found the music an incomprehensible wall of sound and wondered how I was going to stand three hours of it. The expressions on the faces of other guests told the same story. But after perhaps an hour, a strange shift in perception occurred. The music seemed to come in waves, approaching with an insistent rush and an increase in tempo that grew and swelled, filling the listeners with exhilaration. One by one, we rose from our places and started swaying and dancing, faces glowing. The trumpets swung over our heads, the drums crashed, the chanting gathered pace. We began clapping, and moving faster, led by the musicians into an increasingly frenetic dance. Everyone wore ecstatic smiles. It's impossible to think about the petty concerns of life while your brain is being pummelled into submission.

Then the waves of music ebbed and we found ourselves in silence. A silver cup containing milk was brought out and blessed by the head of the brotherhood, then offered to our sick friend to drink while a prayer was said. We closed our eyes for a few moments and willed her to be strong. When the music started again it was as if a switch had been thrown, and the mood swung back to an ecstatic celebration of life.

No one was left unaffected. One previously reserved English-woman who'd been sitting on the edges was now leaping around as though she were plugged into mains power. The rhythm built to its final crashing crescendo, leaving us high with exhilaration. I agreed with Simon – Sufi ceremonies were like a rave party, but with better music.

Sandy and I had a wealth of material from the festival to post on the weblog and turn into newspaper articles, but as we were still waiting for Maroc Telecom to connect us to the Internet, we had to do it all from a local café, where the keyboards were in Arabic. My six or seven visits to Maroc Telecom had resulted in assurances that a technician's arrival was imminent, but none eventuated. It was a bit like *Waiting for Godot.* I was told on one visit that we couldn't possibly have a problem because we weren't on the list of people who had problems. How one made it onto the list was a mystery.

Then Jenny's mother, hearing the story, joined the campaign. In Fez for a holiday, she took herself and Jenny's translator off to lay siege to the head technician in his office. Her petite, birdlike appearance belied her determination: she told the technician she had flown all the way from Australia to sort out this problem, that it was an urgent business matter and she needed it done *now.*

She must have been formidable because the technician got

on the phone and we were online within an hour. In Morocco, personal contact makes all the difference. At last we were able to sit in the comfort of our downstairs salon and file our stories and pictures.

But our battles with bureaucracy didn't end there. Late one night, I opened the door to find a local government official known as the Maqadim, who had visited several times before – once to check the status of our passports, then our ownership of the riad, and again to inspect our *roqsa* and make dire admonitions about changing anything in the house. The Maqadim was a tall, thin bearded man with gold-rimmed glasses and he relished the more officious aspects of his job.

I waited for him to state his business. He wanted to come in, he said in French, to check the work on the house. I reluctantly let him in, saying Sandy was already asleep. Taking him into the kitchen, I pointed out what a good job the builder was doing. He gave the work a cursory glance.

'Where is your architect's attestation to say he is in charge of the job?' the Maqadim asked.

I told him that the original was at the government offices and that I didn't have a copy.

'I must have one,' he said. 'You need to give it to me.'

I couldn't see why. It was self-evident we had one, as we wouldn't have received our *roqsa* without it. If he was so keen on seeing the attestation, couldn't he call into the office? It was right near his. But I didn't want to antagonise him, so I agreed to drop it in.

'When?' he insisted. 'It must be this week.'

It seemed that the Maqadim was intent on hassling us until he found something wrong. He had developed the habit of knocking on our door when he knew our workers wouldn't be around. I suspected he was hoping we'd slip him a handful of dirhams to leave us in peace, but as he never asked and we didn't offer, he kept turning up.

The majority of Moroccan officials we encountered were pleasant and helpful; this man was an exception, and unfortunately this wasn't the last time we were to cross swords with him.

Although Sandy and I were used to spending a large amount of time together in Australia, working side by side for so long was a new experience for us. Mostly we managed to balance the multitude of demands made on us during the day pretty well, still maintaining our sense of humour, but there was a palpable sense of relief when the workers left and we could sit down for a drink and a chat.

Sandy had been making an effort to learn Darija, and during the day he would go around the house every half-hour, checking what people were doing. If this wasn't done, the workers would assume they knew what we wanted, then spend ages doing something that often had to be redone later. Sandy had a sympathetic way of dealing with people that made them feel he was on their side.

'We are not used to having bosses who are so friendly,' Si Mohamed told us after a few weeks. 'Usually we get shouted at.'

Sandy never shouted but nor was he a pushover, and he could be firm when necessary.

While Sandy did most of the supervising, I concentrated on the design side of things – door frames, balustrades, tile patterns, furniture, fittings and finishes. Having researched what was traditional then adapting a design to suit our house, I would communicate with the artisans via Si Mohamed. Sometimes, naturally, they had their own ideas about the way things should be done, but I tried not to lose sight of the overall design. New elements had to blend with the old, rather than compete with them.

I did most of the cooking, and in the absence of a cleaner, Sandy took that task on, an arrangement that sometimes had the locals bemused. He was brandishing our new vacuum cleaner one morning when the workers arrived. They gave him odd looks. I knew that a Moroccan man would never be caught dead doing housework, and thought it was good for them to see the 'grand patron' of the riad taking such things in his stride.

The day before, Sandy had been hanging the washing on the line when a woman on the terrace next door burst out laughing at him. Sure, this was a cultural difference, but to me, women who ridicule men when they do housework are their own worst enemies.

Sandy was given a reprieve when we found a new cleaner, a university economics student who came a couple of times a week. She wore a headscarf and was outwardly traditional, yet she was also a born flirt. Some of the male workers joked about taking her as a second wife.

We had emerged from the cocoon-like environment of the

music festival to have our house invaded by electricians. In our experience of electricians in Australia, they'd mostly worked quietly and been expensive; here they were the reverse. The financial cost was reasonable, the psychic assault anything but. Like army ants on amphetamines, they attacked our beautiful walls with hammers and chisels, shrouding the courtyard in dust through which stone and plaster chips flew like shrapnel.

We had marked the walls in coloured chalk to indicate where we wanted lights, switches and power points. The current wiring had been installed in 1940s conduit on the outsides of the walls. Many of the power points didn't work and the capacity of the system, at 110 volts, was such that you could blow a fuse by running two appliances at once. We had to decide what was most essential at any given time, the fridge or our laptops.

Now these chalk marks became a kind of lunatic join-the-dots game, with muscly young men making deep tracks for the new conduit in our ancient walls, which leaked copious quantities of sand. We tried not to cringe at the channels in walls that were supporting a huge weight above them.

This trench warfare moved so fast we had to frantically shift our belongings from room to room, shoving everything into plastic bags, to keep ahead of them. We had thought we could confine them to the first floor until the lights and switches were installed, after which we could move up and they could move down. But they insisted on attacking the riad on all fronts. Nobody could explain why – it simply was.

By four o'clock on the first day, we felt shell-shocked and

jittery. Our house had become a war zone. It was times like this I wanted to take a holiday from the life we'd chosen. I had to remind myself why it was we'd travelled to the other side of the world at great trouble and expense to live lives our grandparents had happily rejected. The romance of having no hot water, and boiling the big copper kettle for every set of dishes and bucket bath, had soured. I regularly burned my toes as I poured the leaky kettle.

Then there were the hours spent crouched on the floor handwashing clothing, sheets and towels. And yet, compared to almost everyone around us, we were well off. The locals who filled every available container with water at the fountain down the alley and carted it home had no choice; they couldn't afford the water bills. I had to admit I was a spoiled Westerner who found it hard to go backwards in my living standard. I liked having hot water on tap, and above all I liked hot showers. The prospect of doing without for several more months wasn't appealing.

But eventually the trenches were all dug and the electricians moved on to laying the cables, a process which was less dusty but just as chaotic. Like feral spaghetti, cables colonised the entire riad, and walking around became an exercise in avoiding them. Underfoot they threatened to send us tumbling downstairs; they reached out from walls and attempted to ensnare us as we passed.

Several days into the age of electricians, I was aroused from a siesta by yelling. A woman was screaming at Mustapha from the adjoining parapet. He shouted back, others joined in, and the whole thing became a slanging match.

I hurried out. 'What on earth is going on?' I asked Si Mohamed.

'The neighbour says the plaster is falling off her wall from our banging. Mustapha is telling her it is her fault for not letting us inside to check it in the first place.'

It was the same woman who'd told us to go away when we asked to look at what was happening on her side of the wall. Immediate action was called for. Grabbing the nearest suitable gift, which happened to be a bright yellow melon, I collared Mustapha, Si Mohamed and the hapless Jon, who'd chosen that moment to appear, and we marched around to her door.

Presented with the melon, the woman changed completely. As we were taking her seriously, we were honoured guests in her house.

But Mustapha's attempts at explanation didn't seem to be working. His hand gestures became more and more emphatic and I could see the whole thing was in danger of deteriorating. It was at that point Jon's presence became crucial. He turned out to be exceptionally practised at apologising in Darija. We were sorry to be disturbing her family. We were sorry about the plaster. Yes, we would fix it. We were sorry for existing . . .

Somewhat placated, the woman led us around her dar, which had ancient *masharabbia* screens around the balustrade. There was a sense of entering a time warp, with the clock fixed at 1650. Everything was in 'original' condition – old, worn and filthy. Cockroaches were making themselves at home everywhere.

But the house was also charming in its own peculiar way. Climbing level by winding level to the roof terrace, we tried to work out which of our rooms was sited above theirs. As the houses were built on the slope of the hill, their first floor was level with our ground floor. Eventually we identified the back section of our *massreiya* wall, the one that was cracked and bulging dangerously. The cause of the problem was immediately apparent. A chicken coop had been built on the neighbours' roof, and to support it a beam had been plugged into the middle of our wall. Whenever they cleaned the coop, the water drained straight into our wall.

As we were about to leave, Mustapha learned that the house had recently been sold to a Spanish woman, who was taking possession at the end of the month. This was good news – hopefully she would be interested in contributing to a bit of structural work from her side of the wall.

I still had the unfinished business with Hamza and his carpenter to tidy up. They arrived at three one afternoon and did a tour of inspection, Sandy and I following and pointing out places where the workmanship was poor – doors that didn't shut, ill-fitting carved sections, pieces not sanded or finished. The bottoms of several doors had been covered with metal to mask areas that should have been repaired. I could slide my fingers underneath and feel the rotten wood.

I knew the work was of poor quality and suspected that the carpenter, who had a good reputation, had let his apprentice loose on the job. The list of things not done, or done unsatisfactorily, ran to a page and a half.

I doubted that Hamza would have accepted such work in his own house, but we didn't want to fall out with him. The Medina was too small to have enemies, and Sandy and I were prepared to make extraordinary concessions to avoid a dispute.

After the inspection, Hamza went into a huddle with the carpenter, his normally relaxed and friendly demeanour now businesslike. Then he strode across the courtyard.

'The work that's been done is worth fifteen thousand dirhams, so you have five thousand credit,' Hamza said.

Sandy and I looked at each other and gulped. Less than half the work on our list had been done, but by this stage we just wanted the whole sorry saga to be over and agreed to cut our losses. We knew we were being overcharged, but decided to wear it if we could get the three windows made that had been on the list.

Hamza was all smiles and goodwill, conceding that the carpenter could do these windows with the credit we had. We explained where we wanted them, two in the courtyard and one in the kitchen. We shook hands, relieved that things had gone well. Then Hamza said that the carpenter needed another thousand dirhams for wood for the lintels. Still in a jovial mood, we agreed and handed over the cash.

As he was about to walk out the door, Hamza hit us with a bombshell. 'Now you will pay me my twenty per cent commission.'

I was dumbstruck. As he hadn't mentioned it before now, I'd presumed it was included in the total price he'd just given us. From my point of view, we were now being asked to pay an additional four thousand dirhams for work that was shoddy and unfinished, when we'd forgone at least half the work originally agreed to. I was furious, and for once my calm deserted me.

'You've already asked me to pay double the amount for substandard work,' I yelled. 'I am not paying you any more money. You decide – either give us our remaining money back or do the three windows.'

At which point Hamza walked out, leaving me with an expensively learned lesson. My grandmother was right about those who paid in advance. I thought of the old Fassi saying, 'Do not let a rat or a carpenter in your front door', and sighed. Now we had no choice but to look for another carpenter.

The argument with Hamza left me shaken, but the rest of the work continued. As did the unexpected discoveries. Wiring up the downstairs salon, the electricians noticed that the top half of one wall appeared flimsy, and was made of modern bricks. Mustapha climbed a ladder, and wielding a hammer broke through the thin barrier with a resounding crash. He disappeared into the hole and there was a muffled shout.

A moment later he reappeared. 'There is lots of treasure,' he called.

He passed down what looked like a large stone mushroom cap, then another, to a delighted Sandy. They were ancient grinding stones that fitted together, with a hole in the centre for the grain. Next came an umbrella, a set of decorated bellows made of wood and leather, and a cracked plastic baby potty.

'Shall I remove the false wall?' Mustapha asked, hammer ready above his head. Sandy nodded, and within ten minutes the wall was in smithereens on the ground. Behind the false wall was one of ancient bricks, and Mustapha declared them much better quality than the new ones.

So we had now restored the original, and useful, mezzanine at the end of the salon. I loved this organic way of building. No plan? No worries. It was hardly possible to have one when you never knew what lay beneath a plastered surface.

Si Mohamed and I were on the hunt for old *gayzas*, the long thin cedar joists that straddled the ceilings, which were frequently three metres or longer. We needed to replace several rotten ones in the kitchen, and suspected that as we moved into other rooms we would need even more.

We took a taxi to an area outside the Medina, where Si Mohamed suggested we might find the beams, and I found myself in a souk much more rundown and a great deal poorer than any I'd seen. It consisted of a group of concrete-block shacks perched on the edge of an almighty hole – a vast crater perhaps three-quarters of a kilometre across.

'The Hole of Moulay Idriss,' Si Mohamed announced.

Although the crater appeared for all the world like something created by the impact of a giant meteor, it was in fact manmade. This was where the sand and stone had come from when Moulay Idriss II fulfilled his father's vision to build Fez at the beginning of the ninth century. No doubt the hole had been a continuous source for centuries after that, but now it was being slowly refilled with Fez's garbage and building rubble, a process I estimated would take many hundreds of years. Houses teetered on the opposite edge, making it look as though disaster had struck the centre of a city. At the base of the hole, far below the houses, was a series of caves.

From the mounds of garbage rose plumes of smoke, and the scene resembled something from Dante's *Inferno*. Then I saw movement and realised that the garbage was alive. Sheep and goats were picking over the piles, as were people. I watched a boy of about ten walk along a mound, inspecting, discarding, occasionally loading something onto a cart pulled by a broken-down nag.

We headed back into the souk, where amid the shacks a reproduction Louis Quinze bedhead was propped against a wall. Nearby, children were 'restoring' carved wooden chair frames by rubbing them with pieces of broken glass. The *gayza* man Si Mohamed was looking for was nowhere to be found, but it didn't matter as I was quickly distracted by piles of old doors. Although covered in the ubiquitous brown paint, there was a small one that looked promising. Beneath the layers of paint I was pretty sure it was cedar. It was a door that deserved saving, so I bought it – for a fraction of what I would have paid in the Medina.

It took repeated trips to the smouldering Hole of Moulay Idriss before we finally managed to track down some old *gayzas*. They were beautiful joists, a better colour than new timber, and stronger. But they too were covered in several layers of paint.

When we introduced the decapo ladies to the pile of eighty or so *gayzas*, their faces fell. It was an enormous amount of work.

Then Fatima's face lit up. 'Instead of decapo, we will use ammonia,' she proclaimed. 'It will be much quicker.'

Although dubious, I decided to let them clean a sample joist, but the result was disappointing. Not only did it take just as long, but the beam ended up with a furry coat, like a Welsh mountain pony in winter. It was back to the paint stripper, which was more expensive and time-consuming but resulted in an even finish that showed the richness of the natural wood.

As if I wasn't busy enough, in July I decided to go back to school to improve my French. One sunny afternoon I joined hundreds of others in a high-school playground in the Ville Nouvelle, adults, teenagers and children lined up in pairs according to age. In the adults' line were neatly dressed men in suits, men in their twenties wearing jeans, women in djellabas and headscarves, and younger women in tight pedal-pusher pants, clingy tops and crimped hair. The whole gamut of Moroccan society. I was the only foreigner.

At the appointed time, we were marched into a grotty classroom and sat at worn wooden desks covered with years of graffiti. Although it was in French and Arabic, it didn't appear to differ much from the 'Si Mohamed was here' variety. For thousands of years it has been thus – a desire to stamp one's presence on an indifferent universe.

A portly schoolteacher checked our enrolment cards and doled out test papers with rapid-fire instructions – in Darija.

Oh no. I had presumed the language of instruction would be French.

My French was far from flash but it had enabled me to get by to this point. Never one for rote learning, I had leapt in feet first and muddled along. But since arriving in Morocco, my ability had actually deteriorated. Back in Brisbane, our wine-fuelled, Saturday-lunch French conversations with the Belgian teacher had ranged across a variety of topics. Here in Fez, I conversed in only the most basic way, and I was longing for some free and easy communication with Moroccans. I wanted to be able to hold my own in complex and interesting conversations. To do this properly I really needed to speak Darija, but French was a start. All Moroccans learn French at school, and many I met spoke it well.

Sandy had decided to bypass French, reasoning he could always rely on me, and was progressing with his Darija lessons from the workers. His attempts at pronunciation frequently caused them rollicking amusement, and I wondered if they were entertaining themselves with that time-honoured tradition of teaching the novice rude words.

There are some English words that cause Moroccans to guffaw with embarrassment, including 'twenty' and 'zucchini', which evidently sound like Darija names for female genitalia. Conversely there are some Moroccan words I have trouble saying without grinning like a silly schoolkid. These included *Arfuk*, for 'please', and the much used *kunt*, the past tense of the verb 'to be'.

I'd learnt a few Darija words myself, of course, for things relating to the restoration. There were *gayzas* and the *roqsa* and also *traab* (rubble), *zbel* (household rubbish), *hmar* (donkey), *shnou?* (what?), *schuuf* (look), *mezyan* (good), *mezyan bezzaf* (very good), *mumtaz* (fantastic) and *mumkin* (maybe). Our most used expression was *mushi mushkil* — not a problem.

As I stared down at the placement test, I felt overcome with panic. It was like one of those nightmares in which you find yourself resitting your high-school exams. I scanned the page, which offered multiple-choice questions, the first of which was *Si allant au cinéma? Il y a une demande, un imperatif, un ordre ou une interrogation?* Who the hell knew? Certainly not me. I cast a sideways glance at my neighbour, a man from further down the African continent, who appeared to be as stumped as I was.

I took a stab that it was A, and from there on the questions got worse. At the end I had to write a page-long description of 'my city'. Every tense I knew other than the present had flown from my head, and my piece read like the work of a retarded ten-year-old. My concentration wasn't helped by people who had forgotten to turn off their mobiles, or by the constant interruptions of the teacher, who issued streams of incomprehensible instructions.

I handed in my paper just before the hour was up, my only consolation being that I wasn't the last to finish. I saw a younger woman just beginning her description. At least when I ended up in the class for dummies I wouldn't be alone.

But somehow I was put in the second-top class. Had I fluked the exam or was there a mistake? Whatever the reason, I was way out of my depth, and endured a terrifying class every weekday evening for a month. The bell rang at six-thirty p.m. and we all trooped into the classroom like a bunch of schoolkids. If you were late it was too bad, because they locked the gates. The other students in my class were in their twenties and keen to further their job prospects. I felt uncomfortable and out of place as the only foreigner and the only person over the age of thirty.

The teacher was a dynamic and energetic Moroccan woman in her fifties. She had a Socratic teaching method, her intelligent gaze shifting constantly across the sea of faces. You never knew when she was going to launch a question at you like a guided missile. I sat there in pure dread, muttering, 'Not me, please, not me.' The rest of the class gabbled on fluently in French, discussing intricate points of grammar, while I was still coming to grips with what had been said. Although I was acquainted with the various verb forms, we were not yet on speaking terms. And I certainly couldn't address the past and future versions by their first names. I was used to being treated with indulgence in French classes, and being able to clarify points in English, but here that was impossible – neither the teacher nor any of the other students spoke it.

It was a baptism of fire. At the end of a month, my vocabulary had increased significantly but my verbs were still hit-and-miss.

Building rubble had begun to assume a huge role in our lives. It was something we generated in vast amounts as we dug out floors and walls, and disposing of it entailed shovelling it into sacks and having them carted away by teams of donkeys.

One Sunday, we were in the process of interviewing a new

carpenter when one of our regular donkey men came running into the house, shouting in Darija. Si Mohamed asked him what had happened, and the man gave a long, complicated explanation, the gist of which was that his donkeys had been taken.

Evidently the man had been making a delivery of lime and sand, and was about to start loading the rubble when our least favourite official, the Maqadim, turned up. He announced that since we didn't have a permit to put bags of rubble on the street, he was going to confiscate the animals until we paid a fine, and he had taken the six little donkeys off goodness knows where. It was the equivalent of having your car towed.

This was the first we'd heard about needing a permit for rubble. We had a *roqsa*, a building permit, and it seemed self-evident that building would create rubble. Our riad was at the end of an alley and our bags of rubble weren't blocking anyone's access, so what was the problem?

Regardless of logic, the situation was an emergency. Six donkeys had been kidnapped and the donkey man's livelihood was threatened. This was a full-on donkey hostage crisis.

As the donkey team had been subcontracted by our building merchant, it was his responsibility, and Mustapha decided to phone him, shouting down the handpiece as though he didn't trust the telephone to convey the news all the way to Bab Guissa.

'It will be all right,' Mustapha reported at the end of the call. 'The merchant is an important man and knows the Caid. He does a lot of work for him, removing the rubble when houses collapse. He will sort it out.'

The Caid was one of the top officials in Fez and not a man to be messed with. It seemed a big ask that he would turn his attention from important matters to our donkey crisis, and we didn't hold out much hope, but to our surprise, strings were pulled and a couple of heavies were sent to talk to the Maqadim. He caved in, and a few hours later we found the six donkeys standing outside our door.

Not wanting to show his face, the Maqadim had released the donkeys at the bottom of the alley, given them a whack, and they had obediently trotted up to our place. A much relieved donkey man returned to work loading our rubble with a grin from ear to ear.

But he could barely keep up with the rate at which rubble was produced, and to avoid being buried in it we hired a young man to do nothing but sweep, clean and fill bags of it. Our sweeper had a ready smile and an incredibly literal frame of mind, so that if you asked him to clean an area he would remove everything in his path, including the little marble drain covers that kept out rats, cockroaches and flies – and possibly djinns. My plastic slip-on shoes also disappeared in the clean-up.

In order to protect the ancient *zellij* on the stairs to the first floor, Sandy asked the sweeper to cover them in plastic and top this with an inch of plaster, the idea being that we would pull up the whole lot, plastic, plaster and all, when work was completed. Si Mohamed relayed this instruction, and the next thing we knew, the sweeper had covered the stairs in plastic and upended two bags of dry plaster on them. When his error was pointed out he swept up every last speck and, as instructed, mixed in some water.

'Your idea was not so good,' he said a little while later.

'Why not?'

He pointed to three buckets. Each was filled to the brim with plaster — set hard as a rock.

~

The electricians, having been at the riad for just a week, were almost finished. Overall, we'd been impressed with their speed and efficiency, particularly one bearded fellow with glasses whom we nicknamed the professor. He asked detailed questions about what we wanted and made intelligent suggestions. Around his neck he wore a silver chain depicting a figure carrying a ladder.

'Charlie Chaplin,' came the reply when I asked him who it was. 'He is my hero.'

This did make sense, given that The Tramp's movies involved all kinds of gags with ladders, and reflected some of the slapstick situations the professor himself had experienced in his work.

The professor was also a wonderful cook and regularly invited us to share the electricians' lunch, which included things like *kefta* – small spiced meatballs – and eggs. As our contribution we would send the sweeper out to the souk with a pot, which would be brought back filled with *besara*, a simple but delicious split-pea soup. With the addition of black olives and bread, and accompanied by, strangely, Coca-Cola, this made for a more interesting lunch than the sandwiches Sandy and I normally had. We would all sit together in a circle, dipping our bread into the various dishes and making appreciative noises.

David, on one of his inspections, reckoned we had done things in reverse order, and would need to do some rewiring after our structural work. But it was too late to change now. He had never got around to rewiring his own house, and I could see why. He probably couldn't stand the trauma of seeing his beautiful, carefully restored walls burrowed into in such a cavalier fashion.

Fully occupied with running the American Language Center during the week, David usually made his tour of inspection on Saturday mornings. He would come for a cup of coffee and check on the progress. Sandy and I looked forward to these visits and dreaded them at the same time. He invariably found something wrong that we hadn't noticed, and which was time-consuming, expensive and tricky to fix. But we had to admit he was always right.

Mustapha merely dreaded these visits. *'Monsieur David?'* he said, rolling his eyes and flicking his fingers. *'Oh, là là!'*

One Saturday, David took a look in the kitchen, where a hole had been cut in a supporting wall to allow a window onto the courtyard. 'What are you doing? You haven't got a structural support for the roof!' He said this in a tone that made it sound as if we were allowing a three-year-old to play on a busy road. Mustapha's face fell and he scurried off to get scaffolding to fix it.

Meanwhile other things were on the move. After four false and frustrating attempts, we finally engaged a new carpenter. Finding a carpenter in Fez has a lot in common with dating. There's the giddy excitement of thinking you've found someone special, followed by days of chasing them on the phone or waiting for them to ring you. Then he turns out to have so many other offers he can't possibly see you again, or his work is less attractive than you initially thought, or so expensive you can't afford him anyway.

But at last we hired a bright, enthusiastic and funny young man named Noureddine, whose first contribution was to announce that we needed to replace the entire kitchen ceiling. This was made of four huge supporting beams, dozens of *gayzers*, and a layer of planks, revealed when we removed the false ceiling.

We liked the way the ancient cedar beams were blackened with years of smoke from cooking fires, and wanted to keep them as they were. We hadn't allowed for their replacement in our budget at all, but Noureddine thought the ends of the beams looked dodgy, and many were broken. As replacing them meant removing the intricate *zellij* on the floor above, then digging out more than a foot of earth used as insulating material, it was a less than thrilling prospect. But David was also of the opinion it needed to be done.

'The *zellij* is only recent,' he scoffed, '1920s or so.'

So suddenly we were committed to removing and replacing an entire ceiling and floor. As the room above the kitchen led to the one we were now using as our bedroom, I wondered how exactly we would get there.

It also meant another trip to the Hole of Moulay Idriss for more *gayzas*. I was disappointed in the quality on offer this time – they were either bowed or full of rot. Then I spotted some massive beams, the longest almost five metres, as solid as any I'd seen, and so thick I imagined they must have been cut from trees hundreds of years old.

I tried to hide the glow in my eyes, but purchasing them for a good price turned out to be remarkably easy. Getting them home was less so. After being trucked to R'Cif, they were carried to our place, where the porters insisted on much more money than we'd agreed on because they hadn't realised the beams were so heavy. An argument broke out between the porters and Si Mohamed, with a lot of angry gesticulating, shouting, and puffing out and pushing in of chests. The decapo ladies and Mustapha stood muttering darkly that the ringleader was crazy because he smoked too much hashish.

We ended up meeting the porters' demands halfway, and then had to find a way to get the huge beams inside. I figured they were far too long to fit around the turn in the stairs, and then suddenly it dawned on me what one of the old walled-up windows we'd discovered in the kitchen was for. It was in line with the front door, and I'd assumed it had been used for urns of oil and other supplies, but now I saw that it was the perfect size and position for getting long beams in. Within an hour the window was reopened, and with many grunts and the occasional curse, they were pushed through into the house.

Mustapha was ecstatic about the beams. 'These are like gold,' he said. He stroked them, examined them, and even identified which parts of the roof they had come from, according to the nail holes.

There was significant progress elsewhere in the house a couple of days later, with Noureddine completing the restoration of the ceiling in the front entrance, whose planks had been partially eaten away with woodworm. He had installed *gayzers* before replacing the planks, and when the decapo women saw the results of their paint-stripping work in place, there were tears in their eyes. Three coats of paint – white, green and brown – had been stripped off to reveal dark-grained cedar that seemed to glow from within. This small section of ceiling looked suddenly elegant and refined, but made the rest of the house look even shoddier.

Working down my to-do list I went hunting for kitchen items, and had a sobering reminder of the reality of most Moroccans' lives. Si Mohamed and I went up and down Rue Cuny in the Ville Nouvelle trying to find a kitchen sink that was more than six inches deep. When I did find one of European depth, they wanted the equivalent of six hundred dollars for it.

I made the mistake of voicing my thoughts aloud to Si Mohamed. 'In Australia I could buy that much more cheaply. And I'd have a choice of all sorts of designs.'

As soon as the words were out of my mouth, I realised what I must have sounded like. I loathe it when people play the comparison game, dissatisfied with where they are at present. Yet here I was doing exactly that.

Si Mohamed bluntly put me in my place. 'But we are not in Australia. People here have simple kitchens.'

I castigated myself. I remembered having dinner in a fancy restaurant on our first trip to Fez. Going out to the toilet past the kitchen, I glimpsed two women huddled over a single gas-burner on the ground, from which they had produced six amazing courses.

I eventually found a simple ceramic sink that was marginally deeper, and bought it. Next, I needed stone to make bench tops. Moroccan stone comes in beautiful subtle colours – patterned beiges, mottled yellows, creams, pinks, black and green. I really liked a cream-coloured one that looked as though it had translucent grains of fossilised wheat in it. The drawback was that the stone wasn't very hard and scratched easily. The other choice was Italian marble, but I'd seen so many laminex bench tops trying to emulate it that when I saw the real thing it looked somehow fake. I took away a couple of samples of local stone and caught a taxi back to the Medina, dropping Si Mohamed off on the way.

Getting out at R'Cif, I found a large crowd blocking the footpath on either side of the road. Curiously there was no talking or pushing or shouting; everyone was standing still and silent, as though waiting for something. The only sound was a whistle, like that at a basketball game. Peering between people, I saw what had their attention. Lying motionless on the ground was a boy of about five or six, his legs covered in blood. His eyes were wide with shock but he did not whimper or cry out. As I watched he moved his legs.

The astonishing thing was that no one was with the boy, holding his hand or his head, comforting him. He lay apart in his own island of pain. Cars continued to drive around him. There were no hysterical relatives, no one rushing forward to administer first aid. The crowd was simply waiting and watching. I thought about pushing through and going to him, but I felt the force of the crowd like a barrier. I found excuses not to intervene. This was not my drama. What could I do? I couldn't speak his language. I had no medical skills. There was a policeman there, and presumably an ambulance was on its way.

Walking home, I felt ashamed of what I felt to be my complicity in the crowd's passivity. Maybe, I thought, this was due to the belief that events unfolded as Allah willed – you could not do anything to avoid your fate. I was constantly amazed that there weren't more accidents in Fez, given the way people wandered all over the road amid traffic. Today was the first evidence I'd seen that they didn't always do so with impunity.

A few weeks later, I was walking down an alley and was thrilled to spot the same young boy, his leg in a grubby plaster cast, sitting in a doorway. He looked bored and frustrated, but at least he was alive.

Although the electrical work was now finished, the electricity account still hadn't been changed into our names. I went with Si Mohamed back to the utilities office at R'Cif, where for some reason the clerk wanted to see my *roqsa*. We trudged home to fetch it and returned.

'No, not this permit,' he said. 'The other one.'

'Which one? I don't have another.'

'You must have,' he insisted. 'I want to see the one for operating commercial premises.'

'But I'm not operating anything commercial.'

'Then why do you want four lines into your premises instead of the usual two?'

I had no idea why he was asking this when I only wanted to change the name of the account. 'Because that's what the electrician said we needed for the lights, washing machine, fridge, oven and heaters. Two lines won't do if we need to run everything at once.'

The man stared at me as though it had never occurred to him that anyone would contemplate such a thing. He said, 'Then you need a permit saying you are not going to operate commercial premises.'

Somewhere else I would have thought he was making a Kafkaesque jest. But he was serious, and so we headed to the *baladiya*. Thankfully the nice chap was there.

'No,' he told me, 'you don't need another permit.'

'Well, why won't they just put my power on how I want it?'

'The trouble is you've requested four lines. For such a request you need to fill out the appropriate form, pay four thousand dirhams and supply the architect's plans. Then a committee will come and inspect your property to see if it's suitable for commercial purposes.'

'But I don't want to run a commercial operation. It's a private residence – it's simply a big house and I have several appliances and lots of lights.'

'Then you'll need to get the committee to confirm this.'

My brain was dizzy from going round in circles, but suddenly I saw a way out. 'What if I only ask for two lines?'

The nice chap smiled. 'Then you don't need any forms.'

It was a win, but at a cost. I had planned on giving the gas stove I'd bought the year before to Si Mohamed's mother, who didn't have one, and installing an electric oven and a gas stovetop. But now I wouldn't be able to run an electric oven, and probably not a washing machine either. When it grew colder a few months later, we blew the entire electrical system by turning on a single radiator.

In the Medina, people are used to living with just a couple of lights and a single gas burner, so what else would you need modern Western lighting and appliances for, if not to run a guesthouse or a factory? What I wanted to know was, if I was paying for it what difference did it make to the electricity company?

I never discovered the answer to that.

*C*t was now the end of July and the heat beat down into the courtyard in the middle of the day, making the dust stick to sweaty bodies and the work seem much harder. Outside the thick walls of the Medina, it was even hotter. The seasons in Morocco can be extreme. August regularly has days exceeding forty degrees, yet during the past two winters it had been cold enough to snow in Fez itself, something which hadn't been heard of in more than a decade.

One Saturday morning, while David was making his tour of inspection, Mustapha's recurring nightmare worsened. This time David was joined by Rachid and Zina. Things were fine until they saw the wall in the downstairs salon that had been partially removed to reveal the ancient mezzanine.

'But what weight was it carrying?' David asked, leading the delegation upstairs to inspect.

'Monsieur David is always looking for a disaster,' muttered Mustapha, following after him. He might have thought David was overreacting, but a nasty crack in the wall above had grown worse in the past few days and now stretched to the corner of the room.

Rachid and Zina were alarmed, and insisted we shift our bed from its position right next to the wall, where there appeared to be nothing much to prevent the entire floor from caving in and taking us with it. With a thick layer of dirt between the floor and the ceiling below, the weight was substantial and a collapse at one end could lead to a domino effect along the whole length.

Back in the downstairs salon, Mustapha, on instructions from Rachid, removed a *gayza* to reveal a badly rotting supporting beam, and there were calls for immediate scaffolding. A debate was then carried out in Darija, French and English about the whys and wherefores of using metal beams in place of wooden ones to take the weight. Being a purist, David was vehemently against metal.

As I'd managed to buy some magnificent wooden beams just days before, I couldn't see what all the fuss was about. We would use one of them. Plaster was put on a small section of the crack in the upstairs salon, so we could see whether it was still growing.

Checking the strength of the outside wall along the top alley, Rachid discovered a massive old lintel hidden in a cavity over the stairs to the terrace. The door it had once protected was now walled up and invisible under plaster, but he said it was likely to have been the main entrance off the street for that wing of the house.

We paid another visit to the woman in the adjoining house, who was shortly moving out. Rachid gave cries of delight on seeing the ancient wooden *masharabbia* screens that formed the balustrade. It was in such original condition, nothing messed with, that he found it jaw-dropping.

'I am just pleased I saw this before I died,' said Rachid quietly, snapping a few photos.

All the experts were of the opinion, as we'd thought, that the neighbour's house had once been joined with ours. The area where our riad now stood was likely to have been their garden. The bricked-in window we'd found in our kitchen suggested there'd once been a view onto the garden from their place.

There were additional clues. Halfway up our downstairs salon wall was another lintel, suggesting a door through from the neighbour's salon. The most likely explanation for the building of our riad was that the generation who owned the neighbour's house had died and the inheritance was split between several grown children, who needed to expand to accommodate their families.

The two sections of our riad had evidently started off as separate wings with their own entrances, as Rachid's discovery of the lintel of the old door onto the top alley showed. When we started to fix the floor of the upstairs salon, we discovered the remains of a line of green tiles with an exterior finish: this salon, a later addition to the riad, had once been the neighbour's roof terrace, at a time when the whole structure was only one storey high. The *massreiya* would also have been added later.

In contrast to their excitement over the prospective failure of the supporting beam in the salon, Rachid, David and Zina were nonchalant about the chicken coop that intersected the *massreiya* wall, saying the deep cracks were normal and not that difficult to fix, which was a relief.

The workers were removing the last of the insulating layer of earth between the kitchen ceiling and the *massreiya* floor when they found the skull of a cat. I was intrigued. I knew that it had been the practice in many parts of the world to make a sacrifice to the gods of the earth when their domain was intruded upon by construction work: a peace offering was required. Sacrifices were also made to give strength and stability to a building, with animals replacing the original human sacrifices. In parts of Greece, they still kill a rooster or a sheep and let the blood flow onto new foundations.

There is evidence that animals and ritual objects were also used to protect homes in Western cultures. In the 1970s, a curator at the Northampton Museum in the UK realised that the steady stream of people bringing in single old shoes they'd found concealed in their houses was more than coincidental. Many such items have also been discovered during restorations to nineteenth-century buildings in London, and in the Rocks area of Sydney. They were generally found in places where it was thought witches or evil spirits could enter, such as doors or chimneys. Mummified cats, too, have been found in both countries, secreted near front doors.

Ian Evans, an Australian expert on the subject, believes the practice has a long history, a couple of thousand years or more, and that it made its way to Australia with the convicts. It ran parallel to Christianity, with people indulging in folk magic while going to church on Sundays. Since shoes were the only item of clothing to retain the shape of the body when not worn, they were placed in buildings in an attempt to deceive witches and spirits as they roamed the countryside at night, and to distract them from family members.

But no one I asked could tell me if the cat skull in our floor had been put there deliberately. Nor was it our only unnerving discovery. While rebuilding a section of the kitchen, Mustapha unearthed a live snake in the wall. It was a small, blind, metallic-grey thing, and probably lived on woodworm.

Even more unexpected than its discovery was the reaction of the workers, who behaved as if it were a deadly taipan. In reality it was about the size of a chisel, and not acting aggressively. Mustapha wanted to kill it, but Sandy and I intervened. A few days earlier, the workers had caught a lizard and crushed it immediately, claiming it was poisonous. Unconvinced, Sandy looked it up on the Internet and found it was not only harmless, but also rare and endangered. Now he put the snake in a plastic bag and let it go in a dry open drain, where small boys poked it with sticks until it managed to slither away.

Coincidentally, the next day, our neighbour Yusef had an encounter with a much more serious snake. It was a metre and a half long and had appeared out of a disused well in his courtyard. We learned of this when his mother came to our door asking for a bucket of sand and some cement with which to close the well.

Later we found out that Yusef, upon seeing the reptile, had run into the alley yelling, 'Snake! Snake!' Within five minutes, some twenty people had appeared brandishing sticks and bashed the poor thing to death. It seemed amazing to me that even in the most densely populated of urban centres, wild creatures could still make their homes – until discovered by horrified humans.

Yusef was an interesting character and he and Sandy had become friends. He ran a stall in the souk that was little bigger than a telephone box, where he sold spices, beans and coffee. He had a BA in English literature, which he'd studied simply because he liked it. He had a particular fondness for the writings of James Joyce, and this was somehow apt: the squalid yet glorious Dublin of Joyce's imagination has more than a little in common with the Fez Medina.

Yusef's knowledge of English literature had not been of great benefit to him. As many a Western graduate will tell you, it's hard enough for those in English-speaking countries to get a job with such a degree, and in Morocco it was only marginally more useful than a degree in advanced Swahili. After six years, Yusef had given up trying to find a job suited to his qualifications and was resigned to being a shopkeeper. He would have liked to teach English, but the competition to qualify as a teacher was fierce and he couldn't afford the additional years of study. He lived with his aging mother a couple of doors down from us, having stayed to look after her when his four siblings had married.

Their dar was small and plain, tiled with worn *zellij*. It had the simple charm of a house used for living, with little money to spare for decoration. The biggest and most impressive thing in the house was the television, which, we were proudly informed, received two hundred channels. It looked bizarrely anachronistic in the worn old dar.

Sandy and I had been given the grand tour. Most areas of a Moroccan house are usually off limits to all but immediate family, but Yusef led us up tiny steep stairs and through a succession of small rooms, once used for storing olive oil and winter supplies, to his bedroom on the second floor. It had a window out to the courtyard and a serious crack from floor to ceiling in one wall. There was a bare mattress with a blanket and pillow, a few clothes hanging from pegs, a desk and a set of bookshelves. This was the sum total of his worldly goods.

On the bookshelves were academic titles like *Travel, Gender and Imperialism*, along with Conrad's *Heart of Darkness*, Kipling's *Kim* and Joyce's *Ulysses*. Yusef was shy about his own work, but after some prompting showed us a copy of his thesis, titled 'Perceptions of Orientalism in Western Writing'. The level of language was impressive, of a higher standard than most of the essays I'd marked as a university lecturer.

From the terrace we could see a house whose roof had collapsed into the grand salon below. It had happened five or six years earlier, Yusef said, but the family still had no money to repair it.

Over tea, he told us that his distant ancestors had come from an area that is now part of Saudi Arabia. His family had lived in this dar for generations, and both his grandfather and father had worked in the Chouwara tanneries. It was such physically hard work that in winter they would eat four breakfasts, the first at five a.m. and another on the hour thereafter, for respite from the cold. I couldn't imagine what it must be like to spend all of every working day up to your knees in the putrid concoctions used to strip and colour the hides.

Yusef's life was easier than theirs. He started work at his tiny stall around eight a.m., coming home at two for a long nap, then returning at five and staying until midnight. He did this six days a week. There wasn't much room in his life for anything else, including finding a wife.

When David came to dinner a few nights later he walked into the kitchen and stood staring up at the carved and painted ceiling of the *massreiya*, high above. With his thumbs and fingers he made the shape of a square, and squinted through it.

'You know,' he said, 'you could put a *halka* back in here.'

Sandy and I looked at each other. A hole in the middle of the kitchen ceiling so that we could look up and see the beautiful decoration in the *massreiya*. A type of atrium – not a bad idea. It would be surrounded by a balustrade on the floor above, and would also link the two levels of the house, giving the downstairs a sense of grandeur it currently lacked. Furthermore it would return the house to how it once had been.

But I wasn't keen on what we would lose – a lot of floor space in the *massreiya*, which would essentially become a gallery around the four sides of the balustrade.

We slept on it, and next morning it still didn't seem like a completely wacky idea. Not, at any rate, as wacky as David's previous idea, which was to demolish the catwalk joining the two sections of the house, since it had been put in at a later date. As the stairs to the upper salon no longer existed, this would have given us no way of getting there – a minor detail.

We called Zina to get her opinion of the Hole of Amster, as we took to calling the *halka*, and after discussions with David, Mustapha and Rachid, she gave her approval. And so it was decided.

In the meantime a message was relayed from over the wall: I was wanted next door. I walked around to the entrance in the next alley with Si Mohamed, expecting to be told that more plaster had fallen off the neighbour's wall, or that their bedroom roof was sagging from the rubble we'd piled on the floor above it. But instead I was presented with a paper bag, inside which was a pair of elaborately sequinned pink slippers – in my size.

I was amazed and touched. I didn't know what I had done to deserve such a present. Si Mohamed said afterwards it was because I had taken seriously their concerns about plaster falling off their walls from our banging.

The building project had now been going for more than two months and we were concerned that, with roughly a third of our available time and half our budget eaten up, we still hadn't managed to get Mustapha and his men out of the kitchen. They hadn't yet touched the rest of the house – the catwalk that looked in imminent danger of collapse, the roof that leaked every time it rained. Getting the project finished this year was starting to look doubtful, but we couldn't leave the house half finished when we went back to our jobs in Australia.

The more anxious we became, the more it seemed the workers dawdled. If we weren't actively supervising – if Sandy and I went out together for a couple of hours – hardly anything got done. But what did they need to hurry for? As they saw it, when the project was finished they would be out of a job.

We both needed a break, but one of us had to stay and supervise. Sandy obligingly offered to let me go first, so with Jon and Jenny I hired a grand taxi for the day and went to the villages of Sefrou and Azrou in the Middle Atlas Mountains, south-west of Fez. Jon and Jenny were looking for carpets and furniture and I went along for the ride, with a view to buying a carpet too if the right one was to be had.

As we were foreigners, our driver needed to take our passports to a police post before we left town, something that happened on every car journey out of Fez. Then, on the outskirts of the city, we passed two policemen standing vigilantly at the side of the road, checking cars, their occupants and belongings. The heavy police presence on Moroccan roads is one indication of how hard the government is working to avoid further fallout from September 11, and to prevent the infiltration of violence into their normally peaceful society. Many arrests have been made and a number of terrorist plots diverted. In cities and towns all over the country are billboards showing a giant red hand gesturing 'stop', with the words 'Don't touch my country' written in French and in Arabic.

The roots of the violent elements go back to the late 1970s and early 1980s, when the US government was funding and training the Islamic Mujahideen to fight the Russians in Afghanistan. Young

Arab men looking for a cause, or just looking for trouble, were drawn from all over the Middle East and North Africa – among them a charismatic and wealthy young man named Osama Bin Laden. The Moroccans in their number subsequently founded the Moroccan Islamic Combatant Group (*Groupe Islamique Combattant Marocain*), one of the goals of which is to make Morocco an Islamic state. According to *The Independent* newspaper, the bombers who carried out the 2003 suicide bombings in Casablanca that killed twenty-six people, and those responsible for the 2004 attacks on trains in Madrid where 191 died, were under instruction from this group.

Shortly after the invasion of Iraq in 2003, Sandy and I were watching the Arab television station Al Jazeera with a group of Moroccans in a café. As footage of the assault on the city of Falluja was aired, showing the maiming and killing of children, women and old people, some of the men around us grew furious, shouting at the television, hitting their fists into their palms. Others simply covered their faces. I wondered if people in Australia were seeing what we were – what their government had committed them to.

Despite the anger of Moroccans at what was being done to their fellow Muslims with the support of the West, we never felt any of it directed personally at us. Connecting with other people, foreigners included, is a very strong element in Moroccan culture. The Moroccans knew that we were as disturbed by what was happening in Iraq as they were, and they responded to us as fellow human beings rather than as a specific nationality.

Inevitably, though, there are young Moroccan men who continue to answer the Jihadist call. Moroccan authorities say that more than fifty volunteers have gone to Iraq as fighters or suicide bombers, though the real figure may be higher, since not all who make the journey are traced. But the government has cracked down on recruiting networks, including those in Algeria, and continues to make numerous arrests.

~

The journey to Sefrou took a couple of hours, the landscape changing from the Sais Plain to olive groves to rocky fields as we began to ascend the foothills. Unlike much of the rest of Morocco, the vegetation in the Middle Atlas is luxuriant. The region is sparsely populated by Berber people, and villages are few and far between. It felt good to be out in fresh air after so much dust and rubble at home.

As the road began to climb more steeply, we came across orchards, with roadside stalls selling produce. It being mid-summer, the cherries had just finished, but there were other rows of dark trees, their branches weighed down with the golden orbs of oranges. The pickers were in among them, men and women in colourful, practical clothing, some with small children on their backs. Donkeys with panniers waited patiently nearby to carry the loads of fruit.

Sefrou is a charming town, intersected by a rushing river with a picturesque bridge and surrounded by impressive crenellated ramparts. It takes its name from the Berber tribe that originally settled there more than two thousand years ago. That tribe had converted to Judaism, but in his usual persuasive way, Moulay Idriss I ensured they became Islamic in the eighth century. Four centuries later, the town grew wealthy through trade with the Sahara, and a hundred years after that, a large number of Jewish refugees arrived from southern Algeria. Now the majority of Jews have gone to Israel, but the architecture of Sefrou's Medina still reflects their influence, with external balconies overhanging the alleys.

Leaving the taxi outside the main gate, we wandered through the curving streets, looking for the old cedar furniture that we'd been told the residents of Sefrou were discarding in favour of wood veneer. The town had a reputation as one of the best places to buy such things, but although we scoured the streets we didn't find a single item.

There was every variety of fruit and vegetable, of first-rate quality – shiny purple eggplants, bright yellow zucchini flowers, glowing tomatoes, luscious lemons, just-picked plums, gently ripening avocados. If we'd been making a slap-up banquet for fifty we would have been in heaven. But sadly, nothing you could sit on.

Just as we were giving up, we spotted a familiar face. It was the waiter from Café Firdous in Fez, who confirmed that, yes, there was a market on today but it wasn't here. Kindly he offered to show us the way, and we followed him for some distance through the backstreets. Our hopes rose as we came upon a number of parked vans, and beyond them a stream of people heading for a ramshackle clutch of tarpaulins. Surely this must be the market.

It was. But the entire place was filled with more fruit and

vegetables. Crestfallen, we bought a kilo of nectarines and slunk away, all thoughts of French colonial wardrobes, chests of drawers and chairs expurgated from our minds.

Back in the taxi, we headed south-east to Azrou, by way of Ifrane. Winding our way up the mountain roads, we caught the occasional glimpse of a Barbary ape going about its business among the cedar and holm-oak forests. Then, on the outskirts of Ifrane, a curious cultural displacement occurred. Swiss-style chalets with steeply pitched roofs and shutters started to appear at regular intervals, giving the incongruous impression that we'd somehow been whisked off to the Swiss Alps.

The streets of Ifrane itself are broad and clean, with chalets so Alpine you almost expect a cuckoo to pop out of the shutters. They are the legacy of French homesickness during the colonial era, when petty officials created a kind of theme park so they could pretend they were back in Europe on weekends.

In the nearby forest, the last wild Atlas lion in North Africa was gunned down in 1922, a victim of the French predilection for killing for pleasure. The Romans had put a serious dent in their number some two millenniums previously when they exported thousands to kill those pesky Christians (before deciding Christianity wasn't such a bad idea after all). Nowadays the central park in Ifrane sports a concrete statue of the last lion, beside which tourists pose for photos.

The road into the nearby town of Azrou is also lined with chalets, and its European architecture echoes the health-resort town it was during French rule. Being in the middle of a Berber area, it has a reputation for carpets at much better prices than those in Fez, which have passed through several middle-men. The proprietor of one small shop took us to see the bulk of his stock, kept in a house close by. Propping ourselves on the window ledge, we settled in for a session.

Carpet buying is not something to be done in a hurry. There is a certain ritual to be followed. Even if your eye has alighted on the very thing you want, you cannot simply say, 'I want to see that,' then make a deal and be done with it. You must wait while the carpet seller shows you what is clearly his newer, inferior stock, and you should say things like, 'Yes, that's very nice, but it's not right for me.' Eventually he will offer to show you stock that is 'special, just for you'. These carpets are always old and expensive, and you're told they don't make them like this any more. This particularly applies to the very piece you want.

We spent two hours going through almost all of the stock. Some carpets that looked intriguing when rolled up were disappointing once unfurled. We put things we liked into the *mumkin* – maybe – pile, and when it came time to choose, Jenny and I both wanted the same one, cream with lovely bands of embroidery in reds and browns. Carpet viewing with people other than your partner is potentially fraught for this very reason, but Jenny and I managed to resolve the matter peacefully, with her ceding me my first choice and buying a vibrant orange carpet she liked almost as much.

I returned to Fez refreshed, and a few days later I was coming back from the Hole of Moulay Idriss when I ran into David deep in the Medina. He was on his way to inspect a Koranic school that he was paying to have restored, and I went along to have a look. David spent a lot of his own money on such community projects, including street façades, old schools and fountains. The Koranic school was a single room squashed between houses, with a pretty, *masharabbia*-screened window facing the street. Because of David's exacting standards, the restoration had been going on for months, but the timeframe wasn't helped by some of the workers' methods.

Left to his own devices, the master craftsman responsible for restoring the *medluk* on the wall did less than a metre a day, and David had made him speed up. The *medluk* was now cracking in places, something the craftsman blamed on being urged to go too fast, but which David suspected was a result of not allowing enough time for the lime and sand mix to cure. We had decided against using *medluk* in our riad for this reason; we didn't have the space to make a huge pile of it while we waited the month or so it needed to cure.

David and I stopped for a snack at a tiny stall run by a round, jolly young woman and her rake-thin husband. It consisted of a table, a glass cabinet with simple dishes displayed, and a gas burner with a pot. We ate stewed beans, beetroot salad, fried fish and bread, while eyeing a pan full of sizzling oil into which were being dropped crumbed sardines, freshly cut potato chips and other veg-etables. When the husband put a couple of whole capsicums into the pan they exploded, spattering fat far and wide, including a couple of drops on my scalp. Needless to say, they hurt, and when the woman covered her eyes I thought she must have had a direct hit too – but no, she took her hand away and I saw that she was laughing.

After parting company with David, I walked back through the souk. Glancing into the depths of a herbalist's shop, I saw hanging up at the back various animal skins, including some kind of small spotted wildcat. It was no doubt endangered, I thought, and wondered how you overcame hundreds or thousands of years of superstition merely by telling people that putting bits of animals into magic potions was more likely to result in the vanishing of the species than in any cure.

Further along, above another stall, hung a tiny cage, the floor of which was crawling with baby tortoises. Clinging to the bars were two grey and miserable-looking chameleons. I had only ever seen photographs of them before, and stopped to look. Their swivel eyes were desperately trying to project past the cage, as if willing themselves out of its confines. One of them was so skinny that I doubted it was going to last much longer in such conditions. Finding live insects for food didn't seem to be a high priority for the young stallholder.

I had seen young boys in southern Morocco hunting and capturing lizards to sell to just this kind of stall. These chameleons would have been happily minding their own business in their native habitat before being whisked away by small probing fingers, whose owners were eager for a few dirhams. I couldn't bear to see these reptiles in such horrible circumstances and wanted to buy them. But what was I going to do with a couple of chameleons? We couldn't have more pets if we were going to shuttle back and forth between Morocco and Australia. We already had responsibility for three cats on different sides of the planet, including temporary custody of Tigger, Peter and Karen's cat.

I thought of the abundance of kittens on the rubbish heap at the end of our alley. Every day I'd see one or two huddled together among the detritus. Often they'd be gone by the next day and I wondered what became of them.

One morning, I passed a tiny tabby that was mewing pitifully. It had one eye gummed shut and looked incredibly pathetic. I was in an agony of wanting to rescue it, but knew that any help I gave could only be temporary. If I took it home and fed it I couldn't then put it back on the street, it would be too cruel. There was one animal protection society in Fez, but they mostly dealt with donkeys and wouldn't take cats, so what to do?

I'd kept walking, and when I returned that way the tabby was being fed cream cheese by an elderly man kneeling down beside it in the dirt. Then the lovely old baker from the bakery at the end of our alley came along and poured water for it out of a bottle.

Another man stopped and said to me, 'It's the man along that street who's the problem. He lets his cats go on having kittens then puts them on the rubbish pile. It is not right. It says in the Koran that there is a place reserved for cats in Paradise.'

I smiled in agreement. But one cat in the riad was plenty. I wasn't sure what I'd do if Peter and Karen didn't return in time to reclaim Tigger; I couldn't put her back on the street either. As it was I needed to get her fixed before she got pregnant and our cat problem multiplied. Peter had left me some money for the operation, and when I'd rung around the local vets I found the cost ranged from six hundred to fifteen hundred dirhams, which seemed wildly disproportionate to the two hundred dirhams we'd paid for one of our workers to visit the doctor. No wonder there were so many stray cats in Fez if it took more than two weeks of the average wage to have them neutered.

So why should I adopt chameleons over cats? Apart from a childhood fascination with their ability to magically change colour, perhaps because they were an endangered species. They were tugging at me to buy them, but how would I look after them? I didn't want to replace their prison in the shop for yet another cage at the riad.

I toyed with the possibility of finding a good tree in the countryside to put them in, but where? And if I was followed, which would be likely, another set of boys would find them and they'd end up back in the market. I could put them on our citrus trees, where they might find enough insects to survive, but how would they live through the cold winter? And in buying them, wouldn't I simply be encouraging a trade I despised?

Trying to harden my heart by being rational, I walked away, leaving the chameleons to their fate. But the image of them clinging to their bars pursued me all the way home. I told Sandy about them.

'There are thousands of chameleons in cages all over Morocco,' he said. 'You can't rescue them all.'

I felt like Tessa in John Le Carré's *The Constant Gardener*, who says to her husband, 'But these are ones I can rescue.' In her case it was African people. Chameleons may not rate as high on the nobleness scale, but I figured you helped who and what you could, to the best of your ability.

Later I told Si Mohamed about them, who relayed the story to the workers, who no doubt thought I was soft in the head. But it was what they said next that decided us. According to Mustapha, whereas tortoises were usually bred in captivity, kept as pets and fed lettuce leaves, chameleons were more likely to find themselves thrown live into a fire as part of ritual magic. There was a belief that their skin and bones contained a substance that made errant husbands return to their wives, and a woman who suspected her husband of having an affair would lace his food with bits of powdered chameleon.

The thought of the poor creatures being burned alive was enough to win Sandy over, and Si Mohamed and I returned to the stall. For a mere thirty dirhams the chameleons were ours. They clung to one another as we put them into a plastic container. Carrying them down the street, Si Mohamed became a kind of Pied Piper. Small children immediately sensed there was something alive in the translucent box and followed him, wanting a look. One little boy shrieked at the top of his lungs when Si Mohamed took the top off the box to show him.

Back at home, I took the box up to the catwalk and Sandy lifted the chameleons gingerly onto the upper branches of the orange tree, safe from the inquisitive Tigger. They were thoroughly entwined, not keen to let one another go. But the prospect of a branch was too enticing, and eventually they reached out and grabbed the outermost limb. As we watched, their colour changed, morphing from sickly grey to pale green. The fatter one, whom we decided to call Bodiecia, made an immediate break for freedom, disappearing among the leaves. We named the other one Genghis — not that he looked at all fearsome. He was so painfully skinny he resembled an anorexia victim, and I worried that he was too far gone to survive.

Moments later, he almost fell off the branch, just catching himself with his tail. We watched in trepidation. It was a long fall to the courtyard, and the tiles would be unforgiving. But then a small flurry of wind lifted the leaves, and Genghis seemed to wake up to where he was. He swivelled his scaly head around, goggle eyes slowly taking in his surroundings. Perhaps being suddenly released seemed like a dream. He began to stalk delicately along the branch, his curious, prehensile feet wrapping themselves around it, heading for the uppermost leaves.

Next morning, our new house guests were still high in the tree. Their colour had deepened and was now mottled with pink, to match the branches. Genghis's tongue flicked out at lightning speed and landed a fly. His stomach had a gentle bulge; perhaps he'd caught several more insects. I gave a sigh of relief, and the prospect of dying chameleons thudding down around our heads in the courtyard receded.

We ran a campaign on our weblog over the next few weeks and managed to convince several other riad owners to rescue chameleons and install them in their courtyard trees. Sadly, few of them survived the cats, birds and other dangers. Genghis too gave us frequent scares, seemingly mesmerising himself by staring at sunlit walls and then falling some eight metres to the ground. Somehow he always recovered.

13

$\mathcal{A}$t the beginning of August, a minor disaster struck. Mustapha called Sandy and me into the bathroom one morning to point out a trickle of water coming from a broken pipe on the other side of the wall he was repairing. This side of the house was some three metres beneath street level and as we watched, the flow of water grew, filling the room with the odious smell of sewage. In the Fez Medina, the waste-water and sewage pipes are one and the same.

'The flow has increased because all the housewives have just started to prepare lunch for their family,' Mustapha said. 'You must tell the Caid.'

But it being a Sunday, he wasn't at his office. Fatima, one of the decapo ladies, knew where the Maqadim lived and she, Si Mohamed and I set out to find him while Sandy and the workers tried to staunch the flow with sand bags. Given how much grief the Maqadim had given us in the past, not least the donkey hostage crisis, it felt good to be able to hassle him for a change.

The Maqadim's house was a few streets away, and he appeared at our knock wearing his best attire of white djellaba and babouches. Although he didn't seem thrilled about being disturbed, he obligingly came back to the riad to inspect the situation and then rang the Caid at home, who in turn said he would ring the water company.

Moments after the Maqadim had left, the temporary dam gave way. The trickle became a sudden rush, and a small waterfall of foul-smelling liquid poured in faster than our bathroom drain could take it away. The earth on the floor was quickly soaked through and water began to gush into the courtyard. This was at the precise time that Mustapha was upstairs doing his prayers and couldn't be disturbed, so the rest of us, including Fatima and Halima, rushed to grab more bags of sand and create a channel to deflect the water into the drain next to the fountain.

When Mustapha reappeared he wanted to throw bags of sand into the drain further up the street to block it. But this would mean our neighbours' houses being flooded instead, so we weren't keen on the idea.

Then there was a knock at the door, which we'd left open, and a couple of burly fellows from the water company sauntered in, complete with picks, shovels and various extendable poles for shoving down drains. We cheered and whistled like a crowd at a football match. Sandy and I were impressed, and doubted anyone would have turned up so quickly in Australia. To give the Maqadim his due, he had actually done something about the problem.

The men from the water company trekked upstairs, out the back door, and took the cover off the manhole in the upper alley. One of them began shovelling out copious amounts of rubbish – old bottles, plastic bags, bricks, dirt, and lots of worms – which had been blocking the drain for years, forcing the water to find an alternative route by seeping underneath our house.

A small crowd gathered for this piece of street theatre. While one man kept digging, the other pushed a rod down the drain to free up any remaining blockages. As the water started to make its way through the new hole they'd cleared, the flow into the house slowed, but it did not stop entirely. The men promised to return the next day to mend the pipe, and we helped their memories along by giving them a handful of dirhams as a thank you.

The constant flow of tainted water made the house none too pleasant, so we resolved to eat out that night. But there might be a side benefit, we thought. If the number of flies increased in proportion to the stench, the chameleons would get a good feed.

⌣

We received the yearly land tax bill for the riad, made out to a name we didn't recognise. To date, transferring any official document into our names had been far from simple, so bracing for another round with the Moroccan bureaucracy, I took the bill to yet another government office at Batha, together with the house-transfer document as proof of ownership and a photocopy of my passport. I was shunted around to three different people, all of whom read my documents in painstaking detail before declaring they weren't the person to help me. It turned out I wasn't even in the right office and needed to go next door.

There the same thing happened. The man behind the counter read every word of the two-page house-transfer document, then stared at the land tax bill as if he'd never seen one before. As this was clearly out of his league, I was taken to see the manager, who tried very hard to be helpful. He fetched a large book marked *Commune de Fez* and started to search for our house, running his thumb down the columns of names and moving his lips in concentration. When eventually he found our riad a frown flittered across his face.

'Unfortunately, Madame, the name on the tax bill is different to that of the previous owner.'

Evidently in the thirty years they'd lived there, the old couple had never got around to changing the name on the bill to theirs. I was beginning to understand why.

Nor was this the only problem. Because we had two doors on two different streets, we were liable for two lots of tax, despite the fact that both doors led to the same house. The manager made some calculations on a sheet of paper and flashed me a figure of several thousand dirhams, the amount owing. The first portion of tax had been settled at the time of sale, but the second hadn't been paid for at least eight years and was now horrendously large.

'*D'accord*,' I sighed, pulling out my cheque book. But no, it would not do just to pay the bill. His instructions about what I needed to do next became so incomprehensible that I told him I would return the following day with someone who spoke Darija.

Next morning, the manager told Si Mohamed we needed to go to another office in the Ville Nouvelle, where the mess would be sorted out. I could then return to this office and pay the bill. It sounded straightforward enough.

But of course it wasn't. In the Ville Nouvelle, we traipsed upstairs and down, being directed to one office after another, where we would queue only to be told we weren't at the right place. We ended up outside one office where people were flying back out the door so quickly I felt a surge of optimism about the efficiency of whoever was inside. Finally a charming gentleman told us that, because a new system was being implemented, his office was in such disarray it was beyond his ability to help us, or indeed anyone, and I should return the following week. In other words, it was all too hard and please just go away. So I did.

⌣

Work on the riad continued to drag. People arrived later and later and the spirit of enthusiasm that had been there at the beginning was noticeably absent. Sandy decided it was time for a little Western-style management: a team meeting.

When everyone had assembled he gave a stirring speech, via Si Mohamed, about how much we appreciated their work, how important the project was to us, and how, if they did the right thing by us, we would do the right thing by them and help them to find other work once the house was finished.

Sandy's speech had a discernible effect and the pace picked up. It was as though, reminded of the bigger picture, the workers could see that they had a stake in completing the task.

Now that the unexpected river in the main bathroom had been plugged, work there proceeded briskly. The brick wall was almost complete and Mustapha had eliminated the *flambement*. The room was now considerably wider without the protruding bulge and sported a new niche in the end wall, with a light in it.

It wasn't all plain sailing, though; there were a few injuries and illnesses. Fatima stabbed her finger on a sewing needle hidden in the wood of the salon window. It went in so deep she cried. I treated it with antiseptic and we gave her the rest of the day off. Then Halima developed severe tooth pain. A relatively young woman, she had nevertheless lost many of her bottom teeth and now one of the few remaining was in trouble. It seemed a pity to lose it for want of a dental visit, so I booked her in to a French dentist in the Ville Nouvelle. It was the first time she'd ever been to a dentist.

Unfortunately the tooth was broken and had to be pulled out. At least it was done with anaesthetic, instead of being yanked out by some tooth wrangler in the souk, and Halima was back at work the next day, thanking me with a piece of hand-weaving her husband had done.

It was some weeks since I'd seen Ayisha, but one day she arrived on my doorstep desperate and pleading. She had put off writing her final university assignment for so long, eventually rushing it, that she hadn't had time to correct it. Out of kindness, her lecturer had given it back to her and said she had a couple of days to fix it, otherwise he would fail her.

I took one look at it and could see why. She'd had someone type it who clearly didn't speak English. It was as though they'd got bored, or not been able to understand her handwriting, and a lot of sentences petered out mid-phrase.

Ayisha was an articulate, intelligent woman and she'd had months to do this assignment. 'Didn't you read it yourself before you handed it in?' I asked. She tried, she told me, but found it too difficult. Huh?

The essay was supposed to be a critique of what Fez offered in the way of cultural tourism and how it could be developed. There was a wealth of material on the subject, but Ayisha had said months ago that she didn't know how to approach it. So Sandy and I had given her two sessions worth of help with the structure, and suggested relevant people to interview and research sites on the Internet. She hadn't followed up on any of them. Instead she'd used a couple of basic guidebooks borrowed from us, a few interviews with people she'd met in a nightclub, and not much else. Had I been her lecturer, I would have failed her too.

I couldn't stand by and let her miss out on her degree, so I spent twelve hours rewriting the thing, putting in further references and adding footnotes. By the time I'd finished, it wasn't brilliant but it was passable. I wasn't quite sure why I was doing this. It was supposed to be Ayisha's work, and if she failed it would be her own fault, but I knew she had no place to study at home, and getting a degree was her best chance of improving her situation. I was in a position to help her, so why not? And I was her friend, after all.

A couple of weeks later, Ayisha came to tell me she'd passed her degree, expressing undying gratitude and throwing her arms around me in a long hug. I got the feeling I'd been played for a sucker, but at the same time couldn't help admiring the skilful way she'd done it. When shortly afterwards I took a pair of trousers to be hemmed by her father his own gratitude proved stronger than the breach in his relationship with Ayisha. His workspace near their flat was minute, about four metres square. I had seen him there before, sewing at his machine late at night with Ayisha's cat asleep on the counter. (The cat's name, not surprisingly, was Romeo.) At the age of seventy, her father still worked every day to support three grown-up unemployed sons and a daughter who wanted to leave the country the minute she could.

The trousers were delivered to my house by Ayisha in record time, beautifully stitched. She said her father was so grateful to me he would not let me pay. I was embarrassed about this, as I could easily afford to do so and he certainly needed the money, but I couldn't dent his pride by refusing.

Around that time, Ayisha had had an encounter with an African-American from the Bronx. Her long-distance relationship with the English fellow had been going on for about a year at this point, and the more I heard about him, the more he sounded like a dropkick. He veered between swearing undying love for her and muttering jealously about all the men she must be meeting in Fez. He had let her down twice after saying he'd come to visit, and claimed to like the fact that she was 'innocent', meaning, I presumed, virginal. So what would he do when she wasn't? Stop liking her? It sounded to me like he used her as fantasy material, and had no intention of making the relationship real.

After he failed to show the second time, Ayisha was open to other opportunities, and one came along in the form of the American. He was tall, muscular and good-looking, she told me, and he was also Muslim, a big tick in her book.

'We went out to a club in the Ville Nouvelle and I wore something down to here,' Ayisha said, indicating a spot about six inches below her collarbone and giggling at her daring. 'Everyone was staring at me, thinking, Who is she with? He looks so amazing all the girls want to be with him. I felt so free. It was wonderful.'

Then she found out that her new boyfriend was married to an Italian woman who'd just had a baby. He'd spun her a line about how his relationship with his wife was now platonic and he was planning to leave her, but it was difficult because they had a business together. Sure. Maybe it had something to do with the fact that straight after childbirth his wife didn't fancy bonking him?

Ayisha had never been come on to by a married man before, and didn't recognise the game being played. She asked if I thought he would really leave his wife for her. In a word, no, I said. And even if he did, was that the sort of man she wanted to be with? She mulled this over, but I could see the hormones were still flooding her system and she wasn't convinced.

In the middle of August, Sandy left for France by way of Marrakesh. He'd been invited as guest speaker to a literary event at a chateau in the Dordogne, something which sounded unimaginably glamorous and luxurious after the dust and chaos of the riad. We'd initially planned to go together, but as the house required a great deal more supervision than we'd anticipated, we didn't feel confident leaving it for more than a week. Instead I planned to join him in Ireland in two weeks' time, to stay with his daughter and her family.

Renovation is supposed to be one of the most stressful things you can do to a relationship, just down the scale from having a baby and moving house. After two and a half months living and breathing the project – literally – Sandy and I were doing remarkably well. There were the odd moments when our patience with one another was strained, usually when we had differing opinions about the way to do something. I am annoyingly perfectionist and will make people fix things even if they're only a fraction out. That was why David and I got on so well, but as Sandy pointed out, it was also the reason David's house was still not finished after five years. Sandy and I simply didn't have that sort of time, and his attitude is much more, 'Let's just get on with it.' It's a good balance, each of us tempering the other's weaknesses.

Sandy, being an excellent people manager, would constantly check what our workers were doing. I'm more inclined to tell people what I need and let them get on with it, then be disappointed when they haven't done what I wanted and make them redo it. In truth, our employees were probably happiest with my methods, because that way they could see themselves being permanently employed.

The day after Sandy's departure, the smell of something dead pervaded the courtyard. The sweeper spent hours moving things, searching for the source. I thought it might have been the rat I'd spotted the previous week. With unerring instinct it had run into the stove, six of the workers and the cat in hot pursuit. The stove was turned over and every part of it investigated, with no success. I knew that small scurrying visitors would be an ongoing problem until we fixed our drains. Living in a mediaeval city might have its romantic aspects, but *Rattus rattus* as a house guest isn't one of them.

Now the smell seemed to come from everywhere and nowhere at once. Perhaps the rat had got into the drain and drowned in the flood of sewage, I suggested, but the others thought it would take more than that to kill a rat.

I was sitting having a quiet cup of tea after everyone had left for the day when I happened to glance up into the orange tree. There it was. A black plastic bag rotting in the heat, containing the remains of the fish I'd gutted the day before and forgotten to put out with the rubbish that morning. What was the advantage of having a daily garbage collection if I forgot to use it? I put the bag inside another one and into the freezer until the next day, and didn't tell a soul.

Creating the *halka* in the kitchen ceiling was far more complex than doing a regular ceiling, and David had warned us there were only about four carpenters in Fez capable of doing the job. Three of them so far had given us a quote.

The lowest was from a man called Abdul Rahim, who had a brilliant reputation and had worked on the restoration of the Nejarine Museum and the Karaouiyine Mosque. But he was also a bit of a prima donna, and on the day he agreed to begin work there was no sign of him. His phone was switched off and he was uncontactable.

I waited a few days then grew worried. We needed to move on the *halka* as it was holding everything else up. The last name on the list was a master craftsmen called Ahmed, so I paid him a visit. He turned out to be the fellow who'd given the original quote for our carpentry work and then fallen out with Hamza.

Ahmed showed Si Mohamed and me around the riad he was working on, a mansion of a place owned by a French couple. His work was excellent. He was busy, he said, but could get a good craftsman to do the *halka* and supervise the job. We arranged for him to come over at four-thirty and I left feeling relieved.

A little after four, I was astonished to see Abdul Rahim stroll into our courtyard. I had just been mouthing off about his unreliability to Mustapha, and now here he was in front of me, with Ahmed due to arrive any minute. Mustapha stood grinning broadly, hugely entertained by my predicament.

As David said later, promiscuity gets you into trouble. I felt like I was in a French farce. I had no idea how to get rid of Abdul Rahim, and of course Ahmed arrived early, strolling into the courtyard and stopping in bemusement when he caught sight of Abdul Rahim.

They exchanged a few polite words, but Abdul Rahim must have told Ahmed in no uncertain terms to nick off, the job belonged to him, because next thing I knew, Ahmed was scuttling away. Frustratingly, Abdul Rahim still refused to be pinned down and I had the feeling he was toying with us.

Our young carpenter Noureddine, who was doing the smaller work on windows and doors, got a kick out of seeing Ahmed being sent away with his tail between his legs. He told me that, after bringing Ahmed in on another job, Ahmed had convinced the house owners he could do the work alone, ousting Noureddine. Such were the daily dramas of carpenters in Fez.

My suspicion that Abdul Rahim wasn't serious about doing the *halka* was enhanced a few days later when once again he did not turn up as promised. The previous day, we'd finally agreed on a price of eleven thousand dirhams — a far cry from the four thousand he'd initially quoted — and which I then discovered excluded all sorts of things. Si Mohamed made a phone call and was told that Abdul Rahim couldn't start for another three days. It was now more than three weeks since I'd first approached him, at which time he'd said he could start a week later. He knew he had me in a difficult spot. Even if I did manage to find someone else, they wouldn't be able to start immediately either.

In desperation I ran through the other carpenters who had quoted. Why not get one of them to start the other major job, the catwalk? Then if Abdul Rahim didn't turn up again, I could get him to do the *halka* as well. I put this to Mustapha, who warned me that one master carpenter would not work on another's patch.

Without Sandy to share the management, I felt as though I was on a treadmill of responsibilities. There were up to eighteen workers at the riad on some days, and I was the Cecil B. de Mille of the show. I came to appreciate anew how much Sandy had been doing. Things were happening in every corner of the house, demanding careful coordination. Every ten minutes or so, someone would come to ask what I wanted done, or whether I was happy with what they planned to do.

It was impossible to take a break and I developed a severe head cold, even though the temperature was in the high thirties. I craved the luxury of a space of my own that wasn't constantly invaded by workers. Cooking was a feat. The only tap I now had was some six inches off the ground on the other side of the courtyard. Getting to it meant fighting your way through piles of wood, scaffolding and carpenters' equipment, and crossing a dodgy bridge over the sewage trench.

Minor injuries to the workers continued. Fatima, wearing her goggles on top of her head rather than where they were supposed to be, splashed Decapant into one of her eyes. Copious amounts of salty water saved the day. Noureddine presented me with a bloodily bandaged elbow to be dressed. He had come off a motor bike. Mustapha was constantly hitting his thumb with the hammer, and there were frequent stomach ailments.

These practical problems were straightforward to deal with, but there was one worker who caused me to toss and turn all one night, trying to decide what to do. The man we'd hired as a sweeper was, to put it mildly, a sandwich short of a picnic. The only thing he could do without close supervision was cleaning, but as the wages bill had blown out with all the extra workers, a full-time cleaner was a luxury we could no longer afford, especially one who just hung around for most of the day while getting paid as much as those doing the real work.

When the sweeper burnt the wood around the top of the balustrade on the catwalk while attempting to clean some old paint off — despite my showing him three times how to protect it — I got someone else to do it and told the sweeper I could only use him four days a week, adding that I'd understand if he looked for other work. The rest of the crew were working the normal Moroccan six-day week.

The sweeper failed to turn up the following morning and I figured he'd had a better offer. But at ten o'clock I found him sitting in the courtyard, resplendent in his best clothes. He told me he couldn't possibly work only four days a week because he had a wife to support.

Hence the reason for my sleepless night. What would he do if I sacked him? How would he survive? But he'd survived before I gave him a job two months ago, and would have to again when the work here finished. I hadn't offered him a job for life. Why should I be forced to employ someone I didn't need? I wasn't running a charity operation.

This self-justification went round and round in my head, but by morning I felt resolved. When the sweeper turned up next day

I told him I couldn't employ him full-time and he needed to look elsewhere. In a few weeks, when the carpenters had finished and the wages bill was more manageable, things might be different, I said, but in the meantime I would give him a week's severance pay. He accepted this with grace and I detected more than a hint of relief. Si Mohamed said it was a better deal than the sweeper would have expected, having only worked with us for a few weeks.

Shortly after this, during an inspection, Rachid Haloui declared it necessary to get a specialist plumber in to check our link to the main drain. I had a vision of a state-of-the-art plumber turning up with a tiny camera on a tube, so we could see the nefarious netherworld. Of course, it was nothing like that.

The specialist was what was known as a *kwadsee*, a plumber who dealt with public drains. He resembled a gnome, even more so when he discarded his traditional dress and popped on a woolly cap with a pompom. He had brought a bundle of what looked like long twigs but turned out to be pieces of metal. The only other equipment he had was a pick, a chisel and a trowel. We lent him a spade, and in a very short time he had made a deep hole in the street right outside the front door.

All the while he worked, the *kwadsee* kept up a constant stream of chatter. Si Mohamed relayed that we were lucky we got in first that morning because he had also been asked to find a set of gold false teeth, 'worth a million centimes', that had somehow become lost in a drain. Just how this had happened stretched the imagination. Had their wearer been bending over the loo while brushing his teeth? Throwing up? Having an argument in the kitchen, and shouting so loud they popped out of his mouth and went down the drain?

I asked how old the *kwadsee* had been when he started his job and he held his hand half a metre from the ground. As with chimneysweeps in older times, small bodies were better able to get into tiny crevices.

'In the early days,' he said, 'I used to find all sorts of good things, like gold rings and earrings. These days the pickings are lean.'

He went on to complain about the cost of living in the Medina. When he was young everything was so cheap he had done quite well. Nowadays he could barely scrape a living together. I guessed the hundred and fifty dirhams I was being charged for a couple of hours' work was way above what his regular clients paid.

While digging, the *kwadsee* discarded his shoes and I could see him feeling around with his toes to find where the water pipe was. I went back to my writing, and when I popped my head out the door a while later, the hole in the alley had grown huge and all that could be seen of the *kwadsee* was his grey cap bobbing around in the cavity. He looked to be in his natural habitat.

When he pulled himself out I glimpsed an ancient wall in the darkness below. This was the original city drain, centuries old. Our connection to it was a trench, created by bricking up three sides and fitting a cover above, which had now collapsed in places. The *kwadsee* spent most of the day feeding through a sizeable plastic pipe, joining it up at either end. In some ways it was sad – the end of a system that had served the house for three hundred years – but although it was less romantic, plastic was far more practical.

Hopefully it would do the job until the riad underwent its next major restoration in another century or two.

⌣

Sandy might have been gone but I was far from alone when the workers departed each day. I had a host of creatures sharing the riad with me. Every evening the sparrows performed, twittering in the citrus trees. There were hundreds of them. Some would take off just on dusk in a mass, while others would settle down on the catwalk, only to be roused in an indignant flurry when I went along it to bed.

One night, I went to bed to find the room full of freaked-out sparrows, flying around and crashing into the walls. I had left the light on, and perhaps my bedroom looked more appealing than theirs. They eventually settled down, clinging to the new wires hanging above the bed, which didn't yet have a light fitting. As I wrote my nightly journal I watched their rear ends with not a little concern. I left the big doors open and they were gone by the time I woke up.

Tigger was getting very bold and was now climbing halfway up the lemon tree, causing a great deal of anxious chatter among the birds. Fortunately she never got interested in the chameleons, probably due to the glacial speed at which they moved. The chameleons were doing well. They had put on weight and spent their days beneath the ripening oranges, waiting for unwary flies. They were undemanding pets who more than earned their keep.

Mustapha arrived one morning to find Bodiecia clinging to

a work shirt he'd washed and hung out to dry a couple of days before – his wife was away, so he was looking after himself. When he went to put the shirt on he found a kind of oversized brooch attached. Bodiecia was a definite shade of blue, trying to make herself invisible. Mustapha attempted to offload her onto the lemon tree, but instead of gripping the branch she let go and took an unexpected dip in the fountain. She was rescued on a broom handle and emerged hissing in alarm.

As the chameleons' strength recovered so did their desire to venture further afield. While working on the *halka*, Mustapha called me in and pointed to a green speck on the top of the *massreiya* wall. I peered closely and realised it was Genghis, circumambulating the room where the decorated plaster ended and the wood began. At two storeys high, if he fell it would be the human equivalent of a base jump minus the parachute. I watched fascinated as he determinedly made his way around. Unable to cope with the riot of competing colours, he settled for turning himself green with lots of spots.

One day, it seemed that Bodiecia must have ventured too far afield. Perhaps she got fed up with the noise, dust and banging and went off to look for a better tree. Unfortunately there wasn't one within a muezzin's call of the riad, and we never learned what became of her.

I had plenty of other humans for company too, in Sandy's absence. Amanda, our expat friend, was finally having a house-warming, now that Hamza had finished work on her dar. The last time I saw it, the walls had been completely stripped and were being rendered with *haarsh*. Piles of lime and sand were heaped everywhere, and dust was thick in the air. Now the dar looked delightful. It was a small house on three levels, with five bedrooms and three bathrooms. Of the original features, there remained only the gallery balustrade, some exposed beams, and the *zellij* in a couple of rooms. Everything else was new, but in keeping with the style of the house. The work had taken the best part of a year and gone way over budget, but the result was wonderful.

Hamza and Frida were at the housewarming too. I hadn't seen either of them since the day Hamza left our riad in high dudgeon. Knowing they had financial problems due to not being able to open their guesthouse, I hadn't wanted to further trouble them by asking for the money we were owed. But now it seemed that things were finally going their way. They had been granted a guesthouse permit.

'So Suzanna,' Hamza said teasingly, 'you've been fighting with everybody.'

I admired the way he was able to turn himself into 'everybody'. I just shook my head and smiled and congratulated him on his permit. He and Frida were about to leave for Europe, and by the end of the evening we'd all agreed to get together when they returned.

A couple of months later, I received two thousand dirhams from Hamza. But I didn't take up his offer to get his carpenter to finish the work – Noureddine was doing a much better job.

14

Abdul Rahim finally turned up to do the *halka*, which was a huge relief, but instead of the team I thought I was getting, he had only an apprentice to help out, and needed Mustapha's men to assist with the heavy beams. This was annoying, as they were supposed to be finishing the downstairs salon.

I now had two carpenters onsite, both needing new wood — three thousand dirhams' worth, to be exact. Until now I had bought second-hand wood and had it remilled, or got the decapo ladies to strip it, but the *halka* required longer lengths and a more consistent quality than recycled wood offered.

I rustled up the money, which was no mean feat as it was also payday. I even gave Si Mohamed and Abdul Rahim an extra five hundred dirhams in case it turned out to be more. I couldn't go with them to buy the wood since I had to go to the *baladiya* to renew our *roqsa*.

While I was there, Si Mohamed called to say that Abdul Rahim had chosen some beautiful wood. 'It will cost seven thousand dirhams,' he said.

'Wait right there,' I said, and found a taxi and headed for the wood shop in Bab Guissa.

As I suspected, Abdul Rahim had picked the very best quality wood – and who could blame him? It was just a pity I didn't have bottomless pockets. I asked to see the second quality and was shown pieces with splits in them. Were there longer ones that could be cut down? I persisted. Going through the pile, we managed to choose pieces that were still beautiful but had a few knots and lines here and there.

But at the end of it, the bill had been reduced by only a thousand dirhams. The price of wood seemed to fluctuate from week to week, depending on the supply, and this week, second-quality wood was a little over nine thousand dirhams per cubic metre.

I then proceeded to drive the woman in the shop crazy by checking her figures with a calculator. I rang other expats to see what they had paid, until the woman was ready to brain me. Eventually I paid. It hurt but I did it.

Afterwards, I wished I had bought the first-quality wood. In Australian terms, the price difference was negligible, but here it was huge. And there were so many other things to pay for.

Later that afternoon, I found myself four hundred dirhams short for the wages and had to borrow from Si Mohamed to make up the difference. He didn't hesitate when I asked him. I loved that about my Moroccan workers and friends – they would do anything to help. Still, the notion of a Moroccan worker lending money to a Western boss must have amused him.

It seemed to be my day for unexpected expenses. In order to renew the *roqsa*, a document was needed from our engineer certifying that she took legal responsibility for the safety of the riad's structure while the *halka* and catwalk were being built. I went to Zina's office in the Ville Nouvelle, where she told me her fee for the document was four thousand dirhams. This was a special price, I should understand, because I was her client.

My mouth opened and I gulped a couple of times. Six hundred dollars for a single piece of paper? Rachid hadn't charged me an additional centime for his *attestation* document.

'For a private residence?' I queried.

She changed the figure she'd scribbled in front of her to three thousand dirhams. 'I must pay insurance, you know. It's my risk, not Rachid Haloui's.'

What she means, I thought, is that if the house falls down I can sue her insurance company. To whom she'd have to pay premiums anyway. I nodded, smiling with as much grace as I could muster, chatted on for a few moments about the house, then took my expensive piece of paper and left. Not a bad salary for five minutes' work.

Back at the *baladiya*, I found that, naturally, Zina's document wasn't enough, and I had to send her a registered letter informing her that work had started on the riad. As this was my second *roqsa* and she'd already been working for me for two months, this seemed just a trifle absurd. But I went dutifully to the post office with the letter, then paid a further two hundred and fifty dirhams at the *baladiya* and was handed a new permit. It had that day's date on it, rather than starting when the existing permit expired, as I'd requested. This meant I was being shortchanged by three weeks.

I had a raging argument in French with the head inspector, who claimed that since I'd now paid, it was too late to change it. I pointed out that I'd only been given the date after I'd paid and he tried a different tack, asking when work had started on my kitchen ceiling.

We then had a conversation neither of us understood properly, and he resorted to saying that if our restoration went a couple of weeks past the *roqsa*'s expiry date, he'd overlook it.

I gave in, thanked him and tried to shake hands on our deal. But he wasn't keen, which made me suspicious about how binding his reassurance was. I went home exhausted. These fights with bureaucracy were endless.

The bills didn't end there. Having found a vet who would spay Tigger for six hundred dirhams, I arranged an appointment for a time when Si Mohamed could come with me, being all too familiar with the extraordinary strength of panicked cats. We managed to get Tigger into a cardboard box and set off in a taxi, but I had written down the vet's address incorrectly and we ended up driving around to three different vets trying to find the one I had an appointment with, while Tigger grew increasingly hysterical. Finally we found it, a small, clean-looking surgery with a friendly vet.

I left Tigger to her fate, returning that evening to collect her.

I was instructed by the vet to cover the cut in her abdomen with antiseptic cream a couple of times a day, then spray it with iodine. He also sold me some antibiotics and worm tablets, as it appeared she was infested. The total package came to eight hundred dirhams.

The workers shook their heads in amazement on hearing that so much money had been spent on a cat. They couldn't afford that amount for their children's medical bills or for their teeth — things that mattered. Which is not to say the men were indifferent to the wellbeing of animals. One day, I glanced through the window into the courtyard and saw one of them put down his hammer and chisel and pick up Tigger. He put her on his shoulder and stroked her, to which she responded with great pleasure. It made me warm to this quiet, big man from the Sahara who worked with such silent intensity on our house.

By the start of September, the *halka* was looking fabulous. The kitchen floor had been dug up and the last of the new drains installed. Coming into the kitchen one day, I looked up and caught my breath on seeing the beautiful carved ceiling of the *massreiya* nine metres above, framed in the *halka*. Abdul Rahim's apprentice was doing an excellent job of the decorative facing inside the *halka*, reflecting the design on the beams above the pillars in the courtyard. It was wonderful to see these traditional skills surviving.

The only thing needed to complete the *halka* was the balustrade.

I'd decided to have this made from wrought iron framed in wood, similar to that on the catwalk, with design elements taken from the window grilles in the *massreiya*.

Abdul Rahim and I had also come to an agreement on a reasonable price for the catwalk, and Noureddine too was hard at work. He had pulled out an ancient door from the *massreiya*, frame and all, and found it was hand-adzed all the way around. It probably dated from when the house was built.

Things had reached a stage where it was a delight to wander through the riad after the workers left, savouring what had been done. I started to feel a thrill of anticipation for the time when it would all be finished.

There were also moments of beauty during the busy work day. Going upstairs one morning, I glanced into the *massreiya* and saw Si Mohamed and the apprentice beginning to pray. They were standing side by side, poised on the edge of their mats, their eyes closed in a moment of contemplation. There was something lovely about the way they took time out of their day to reflect, give thanks, just be in the moment – to be reminded that there was something beyond themselves and their individual concerns.

Before this extended time in Fez, I had not understood the extent to which Islam was woven into the thread of everyday life. Calls to prayer rang constantly over the rooftops and the devout among our workers would race down to the mosque, or find a quiet spot in the riad to pray. Mustapha and a couple of others had dark circular marks on their foreheads from a lifetime of pressing down on prayer mats.

At the heart of Islam are the five tenets that every Muslim is expected to follow: accepting that 'there is no god but Allah, and Muhammad is his prophet'; praying five times a day, the exact times of which are determined by the position of the sun; giving alms to those in need; fasting during Ramadan; and making a pilgrimage to Mecca at least once in a lifetime.

As with practitioners of other religions, Muslims follow these tenets to varying degrees. Nearly all our workers, though, seemed remarkably pious. Even the younger ones would talk in great depth about characters in the Koran (many of whom also appear in the Bible and the Torah), as though they were real people whose habits and foibles they knew well. It was a bit like office workers in Western countries chatting about characters in a soap opera around the water cooler.

One day, when they were laughing over something, I asked for a translation and was told they'd been discussing Joseph of Nazareth – how handsome he was, how women would see him and swoon. Because Joseph was a carpenter, Noureddine announced he wanted to be just like him. This was the same Joseph, the father of Jesus, whom I had thought a dry old stick when I learnt about him at Sunday School. Muslim religious lessons were obviously a whole lot more lively.

Islam has a great deal in common with the other Abrahamic faiths, Christianity and Judaism, all three coming from shared traditions. The Muslims respect Jesus as a wise and clever prophet; they just don't believe he was literally the son of god, which to them defies logic. It seems crazy that people have been killing each other over such apparently minor differences for centuries. But as a sixteen-year-old Muslim girl told me, 'The reason people hate and kill one another is because of cultural and political differences. They use religion as an excuse.'

The day before leaving to join Sandy in Ireland, I wrote a long list of things to be done on the house and gave it to Jon and Jenny, who were looking after things in our absence. Sandy often teased me for having 'a touch of the Henrys' – my father too is a supreme list-maker – but it seems perfectly sensible to me.

Although I'd been feeling I couldn't possibly leave Fez for three weeks, now I was looking forward to it. Sandy and I would have a week together before he returned to look after the house and I continued travelling. It would be a relief to get away from the intensity, from being constantly surrounded by so many people, and from the relentless need to stay on top of everything.

The least expensive flights to Europe left from Marrakesh, and I arrived in mid-September after a nine-hour train journey. Not having seen the city in four years, I was struck by the increased traffic. The usually magnificent Koutoubia Mosque struggled into view through a haze of exhaust fumes, and the road was bordered on either side by massive hotels that seemed to have sprung out of the ground fully-fledged.

The taxi dropped me near the Djemaa al-Fna, the famous central square where snakecharmers and storytellers vie with food stalls for the local and tourist dirham, and I spent a bewildering few minutes trying to orientate myself. Marrakesh has double Fez's population and many more tourists, and it seemed like most of them were trying to cram into the square. After a couple of false starts, I found my hotel, left my luggage, and wandered back into the night in search of something to eat.

Djemaa al-Fna too had undergone a transformation in the intervening years. The bitumen in the vast square had been replaced with paving stones, and the food stalls that were wheeled out every night, once higgledy-piggledy, had been tidied up. They were now in neat rows and all had exactly the same kind of wrought-iron frames with shadecloths. They were numbered and lit up, their menus displayed on boards, and the place had the air of a food court in a shopping mall. Where was the glorious panoply of culinary choice in a charmingly run-down setting that had once existed?

Strolling about, I noticed that the prices had been hiked up accordingly. Although you could still get a bowl of *harira* soup for five dirhams, anything with meat or fish was now more than twenty-five dirhams for a tiny saucer. There was also much less variety – it was now mostly soup, salads, kebabs and sheep heads.

I ended up eating salad and calamari, squeezed in between other tourists, then headed for the souk, which at eight o'clock was still buzzing. But here too things had changed. The stalls had many more mass-produced items – Chinese shoes, T-shirts, plastic geegaws. Those that were still handmade were of a markedly poorer quality.

Further into the souk, I was dismayed to see a modern,

fluoro-lit boutique between the stalls. It seemed only a matter of time before this market would resemble those in any tourist city, with identical shops and the same factory goods. And now too there were motorbike hoons – Moroccan boys in tight T-shirts roaring through the Medina and expecting pedestrians to leap out of their way. Why were they permitted to drive through a pedestrian thoroughfare? I wondered. I didn't remember them being a problem last time I visited.

It seems that in catering to tourists, cities shape themselves according to Western cultural values, and in so doing destroy their individuality, the very uniqueness tourists have been seeking in the first place. Mass tourism sanitises a place, making it appear like everywhere else, then moves on in the relentless search for somewhere different, where the same thing happens all over again.

Marrakesh had become like a giant holiday resort, a place where people came for a touch of the exotic safely packaged in a familiar wrapping. It made me fear for the future of Fez. In a voracious wave, the tourists were coming. Cheap flights from Europe were about to start.

Yet who was I to deny local people the money and opportunity that modernisation brought? I only hoped that people would learn from what had happened in other places, and would value and preserve their heritage, rather than sacrificing it to the gods of progress-at-any-price.

Sandy's time away could not have been a greater contrast to life at our riad. He had been staying in pure luxury in the south of France, and his appearance as guest speaker had gone over exceptionally well. Then it had been on to Ireland to stay with his daughter and her family, where there were taps that worked, clean chairs to sit on, and a functioning kitchen.

I joined him there, and found being surrounded by people who spoke English bliss. I didn't have to call for Si Mohamed, or struggle to describe something in a mix of French, Darija and charades.

When our domestic hiatus was up Sandy flew back to Morocco to take over from Jon and Jenny, and I continued on to Granada in Spain to fulfil a long-held ambition to see the Alhambra Palace, many of whose artisans had ended up in Fez. The construction of the Alhambra, whose name means 'red castle', was begun under the Moorish Nasrid dynasty in the fourteenth century. It is the most significant surviving example of Muslim architecture in Europe.

The city of Granada has a population of some quarter of a million. It is a mixture of glorious Spanish baroque architecture and ghastly, late-twentieth-century blocks of flats. The effect is jarring, like a mouth of perfect teeth spoiled by a few badly fitting ones. Along the backstreets, though, some old buildings are being refurbished rather than knocked down.

On the top of Sabika Hill in the centre of the city sits the Alhambra, spread over some 142 000 square metres. A mélange of buildings from different periods, it creates an unlikely but harmonious whole that dominates the skyline. Once inside the site, it's a long walk to the Nasrid palaces, the main attraction. These palaces were built to display the power that Muslim rulers still enjoyed after being forced to retreat to Granada, their last stronghold. The antechamber, the Mexuar, is a gentle introduction to a series of elaborately decorated salons and courtyards, each more magnificent than the last.

Following the principles of Islamic design, in the centre of each courtyard is either a fountain or a channel of water, creating a unifying element. The intricacy of the plasterwork and *zellij* decorating the rooms surrounding the courtyards is breathtaking.

In the private quarters, the Court of the Lions, colonnades of carved plaster enclose a central fountain supported by twelve leonine statues. A large room to one side of the court has a gigantic star-shaped design in the ceiling, the delicacy of which defies belief; it looks like a multitude of miraculously arranged mini-stalactites, carved out of plaster. This was the room where the last Muslim ruler of Granada, Boabdil, is supposed to have invited to dinner a family with which he'd fallen out, then had them killed.

Most of the palaces were built over a period of thirty years. I pictured the workers living together, sharing meals, having feuds and money problems, speculating on palace intrigues and wars, all the while working on their few centimetres a day. Did they have some sense that what they were doing would extend so far beyond the bounds of their lives? They could hardly have envisaged that five hundred years later, people of diverse cultures, faiths and religions would come from all over the planet to see their work.

Boabdil surrendered to the Spanish rulers Ferdinand and Isabella in 1492, after which the fate of the Alhambra was shaky for a few centuries. Subsequent generations of Christian rulers put their stamp on it, demolishing sections and rebuilding them in styles they preferred. Ferdinand and Isabella are variously remembered for winning Granada back from the Moors after an exhausting ten-year war, despatching Christopher Columbus to the New World, and starting the Spanish Inquisition.

In 1812, Napoleon's troops attempted to blow up the Alhambra while decamping. The fact that the palace still exists today is due to a soldier who made sure the explosives didn't detonate. Only a wall was destroyed, but the site was then left to fall into decay. Nineteenth-century drawings show the wonderful Court of the Lions with holes in the paving, and weeds growing through the marble. Ironically it was the Alhambra's increasing popularity with tourists that saved it. I wondered how much similar, little-known and neglected architecture existed in Fez.

There is further irony in the fact that the Spanish Christians, having spent eight hundred years trying to kick the Moors out of the country, now make so much income from their legacy. Granada is such a Moorish space that it seems strange never to hear the call to prayer.

15

I returned to Fez feeling as if I'd been away for months. I'd had pangs of homesickness, despite relishing the comforts and wonders around me. And even though the prospect of private space was one of the reasons I'd looked forward to leaving the riad, I missed the workers too. I had grown more used to their company than I thought.

Mustapha and the crew seemed genuinely delighted to see me back, filling me in on what they'd been up to. They'd had a paid holiday and all gone to the public swimming pool together, where Mustapha managed to lose his false teeth underwater. I was hardly surprised, as he rarely stopped talking. He had to get the lifeguard to don his goggles and dive to the bottom to retrieve them, causing untold hilarity among the others.

On the down side, though, not a lot had been done on the house while we were away. A major beam had been put into place in the ceiling of the downstairs salon, and the walls had been stripped and coated with *haarsh* in preparation for plastering; some lintels and a couple of windows had been installed in the bathroom, and more *gayzas* had been purchased. But the catwalk, which I had expected to be half finished by now, hadn't been started.

Nor had any headway been made with the plumbing: the shower, two handbasins and another toilet still had to be put in. Jon and Jenny had rung repeatedly to see why the fundamentalist plumber wasn't showing up to work, and heard a mountain of excuses: he hadn't been able to get the right pipes, he had been sick, some relative or other had passed away. By the time I got home, Jenny estimated that four of his close relatives were now dead.

We got Si Mohamed to ring and say we were about to take the job away from him if he didn't get on with it. Good plumbers being harder to find than carpenters, this was a bluff, and he knew it. He turned up for half an hour, walked around with his hands behind his back, then left saying he'd return next morning. He didn't, so we rang and were told he'd arrive within half an hour. Two hours later, after we'd rung once more, he finally put in an appearance. Some pipes actually went into holes and we began to feel hopeful.

But he failed to show the following morning and Si Mohamed was told that yet another relative had died. Really, the plumber was a most unfortunate fellow. His family were dropping like flies.

A couple of days after returning, I had a fully-fledged panic attack, brought on by going to the bank and discovering that expenses while we were away had been higher than anticipated and there was only sixteen thousand dirhams left in the account. Sixteen thousand of anything sounds like a lot, but in reality it was a couple of thousand dollars, with which we had to finish the entire house. We needed closer to twenty thousand dollars, and were forced to apply for extensions on our credit-card limits. They say that owning a boat is like standing under a cold shower and ripping up hundred-dollar notes – restoring a house in Fez is the same, except that our shower was still a bucket.

Sandy and I had a discussion that stretched into the night, focusing on all the things we hadn't done, or should have done differently. We decided we'd been far too nice, and hadn't pushed some people enough. Noureddine, for example, had managed to make repairing an old door stretch out to five days, when there were far more important things that needed doing, such as the kitchen windows. I'd asked him a couple of times when he was going to get onto them; he'd assured me he was just about to and then not done it. It was my fault for not having been more insistent.

While Sandy eventually fell asleep I lay awake worrying until the small hours, feeling an increasing sense of desperation. We had to finish what we'd started, although it would mean racking up the credit cards. To complicate matters, we had Sandy's daughter and grandchildren arriving in a few weeks not nearly enough time to complete everything, especially when trying to get the plumber to do anything was so damn difficult. After three months of solid work, we still didn't have a shower, kitchen, or a dust-free space in which to sleep. We'd anticipated having time to enjoy the house a little before returning to

Brisbane. Now it looked like we'd be working right up until our departure, and even then it might not be finished.

I woke up feeling wrung out, while Sandy seemed a lot chirpier. Downstairs, I skolled the coffee he made me and began ordering people about, trying to convey the sense of urgency I felt. I told Noureddine that if he did the windows well and quickly, then I would retain him to do the cupboards. Otherwise I'd find another carpenter to do them. I had a long talk with Mustapha. He was eminently practical and said he was quite capable of installing the major pipes through the courtyard and making the catch pits they fed into. But we still needed the plumber to connect the other toilet and taps in the main bathroom.

After several more pleading phone calls, the plumber agreed to come the next morning. He was only half an hour late, which was a good start, but after wandering around looking at things for a while, he said he needed to go home to change his clothes. Three hours later, he still hadn't returned.

To make matters worse, the Maqadim turned up again, rapping on the door and muttering darkly about being told we were using modern bricks, firing them ourselves. He said this with absolute authority, despite having just walked past a stack of handmade traditional bricks in the entrance corridor. Not only that, he insisted we were using steel beams instead of wooden ones.

Unable to locate a single modern brick or steel beam on the premises, he looked around for something else to harass us about. In the kitchen, he looked up at the ceiling and almost wet himself with excitement.

'Why is the *halka* there?' he demanded. 'It was not on the plan.'

'What plan?' I countered. 'As we're not opening a guesthouse, we've never been asked to submit one.'

He took Mustapha aside and heavied him. Mustapha said he had no idea what it was all about, he wasn't the boss. When the Maqadim turned his attention to the increasingly uncomfortable Si Mohamed, I piped up and said that the *roqsa* committee had already inspected the *halka* and didn't have a problem with it.

'You have a new *roqsa*?' the Maqadim asked, clearly disappointed.

With a flash of insight, I realised his appearance now wasn't a coincidence. This was the week the old *roqsa* had been due to run out. If there was a prospect of it not being renewed, perhaps we might have been amenable to greasing his palm.

I gave him a copy of the engineer's report and promised him a copy of the *roqsa* later in the week. This didn't satisfy him; he said he'd be making further enquiries and would need to discuss it with us at another time.

The following Friday, our workers' day off, we heard the familiar officious rap on the door. We sat quietly inside, declining to answer it.

A week later, he turned up with two members of the *roqsa* committee in tow. Clearly unimpressed at being dragged back to the same property twice, they had a cursory glance around before departing. The Maqadim went off looking for other people to hassle, and we didn't hear from him for a while.

In the midst of these woes, something wonderful happened.

Sandy had got one of the men to chip away the remaining bulge in the bathroom wall, convinced it was caused not by earth movement but by something more interesting. The old bricks and mortar were removed to reveal a tall urn without a base, inset into the wall. It was the site of a natural mineral-water spring, something houses in this area had also had before the water table dropped at the beginning of the 1970s. We were as excited as Howard Carter must have been uncovering Tutankhamen's tomb.

'Maybe you will find treasure, which you must share with all of us,' called Fatima, who was stripping paint off the ceiling nearby.

Si Mohamed explained that people used to hide their worldly wealth in such places, but as far as we were concerned, the remains of the ancient spring itself was the treasure. Since it no longer functioned, we decided to put a copper bowl on top of the urn, install a tap at the back, and use it as a hand basin, leaving it otherwise untouched apart from a few repairs. This was a much neater solution than a having a hand basin taking up floor space in the narrow bathroom. It was days like these that made working on the house pure joy.

And the workers still seemed to be happy. Mustapha would break out singing on occasion in a deep rich baritone that made us all smile. The men laughed and joked, teased one another, made animal noises. One of them could do a dog howling that made Tigger head for the terrace. Another did a mean rooster. Someone else a cow. The cacophony transformed our riad into an hysterical *Animal Farm*.

Animal noises aside, with our hodge-podge of languages

and gestures we communicated remarkably well, even when Si Mohamed wasn't present.

'I really like you,' Fatima told me one day, apropos of nothing at all. 'You are my sister.'

Perhaps it was a pre-emptive move, the cynical part of me speculated, as the decapo ladies' work was coming to a close. They did not want to leave and their pace had slowed considerably, even though we'd promised to find them other work. Sandy and I didn't really want them to go either; we enjoyed having them around and did not push them to work faster. No wonder our budget was shot to pieces.

Not long afterwards, Fatima and Halima had some kind of argument. This was unusual for them; they were related by marriage and spent a great deal of time with each other, arriving at work together and chattering away happily. Their quarrel changed the atmosphere in the house, and Sandy and I didn't like it one bit.

One morning, they turned up separately, still not speaking to each other, and Fatima refused to work with Halima. I took them aside, and with Si Mohamed translating gave them a pep talk. I began by saying we were very happy with their work, but that to Sandy and me, a home wasn't just about the bricks and mortar, it was about the feeling in the house. At which Fatima interrupted to say it was the same for her, she was working from her heart. I went on to say how unusual it was to find women in the construction industry, and it really helped that they had the support of each other. I knew there was a problem between them but they needed to put aside their differences and work together. Perhaps when they were feeling calmer and less emotional they could talk about whatever it was.

By the time Si Mohamed had finished translating, they had tears in their eyes. Halima embraced Fatima; I gave each of them a kiss and went away, leaving them to it. When I came back I heard the rise and fall of their normal chatter. I never learnt the cause of their argument, which apparently remained unresolved, but at least they were civil to one another and ate lunch together, although they continued to arrive and leave separately.

Another morning, Fatima had a stand-up fight with Noureddine, who'd been teasing her about something too close to the bone. Fatima was screaming and crying and both of them had to be held back from doing one another physical damage. But as with most arguments I'd witnessed in Morocco, it was all over in five minutes. They retreated to their separate corners and sulked. Later, Noureddine apologised by bringing Fatima a cup of tea and some cake, and an uneasy truce reigned.

The last time I'd seen Ayisha she'd been on a high, but she was looking more serious when I ran into her in the street one day. She stunned me by saying in hushed tones that she'd just been to visit a witch. Was this the same woman who rejected all the old superstitions as 'nonsense'? (Before, it must be said, going on to reveal she had once seen a djinn.)

'So what did the witch say?' I quizzed her.

'Nothing. She was no use at all.' Ayisha threw her hands up in disgust. 'I went to ask her what should I do about the two men in my life and she couldn't tell me much. Just general stuff.'

'Did you tell her specifically about the two men?'

'Of course not. It is her job to tell me.'

As Ayisha's mother was waiting for her on the corner, she promised to come around and tell me more in an hour or so.

It transpired that her mother had convinced her to see the witch, since Ayisha was confused about the direction of her life. Show me a 23-year-old who isn't, I thought.

The witch had begun by asking Ayisha if she had any metal in her pockets, metal apparently being a no-no when you're trying to hone in on someone's psyche. She held Ayisha's hand over an incense burner that was wafting out clouds of smoke.

'No one wishes you any ill will,' the witch had told Ayisha. 'Your energy is clear. However, I sense something else. Something more ominous. You are beautiful and a djinn has become jealous of you and is possessing you. There are two men who love you, but you will not marry either of them, or anyone else, because the djinn will not let you.'

Ayisha was laughing as she told me this. 'I went home and told my mother and she was horrified.'

I understood why. For a Moroccan girl not to marry was considered a tragedy of immense proportions, as it meant she could not have children, and in traditional Moroccan society this was seen as a woman's primary purpose in life.

'I don't believe in all that magic rubbish,' declared Ayisha, and in the next breath was offering to come and burn incense and perform a ritual in our house when the building was finished.

While we'd been talking we'd been preparing dinner together. Ayisha was very precise about the way things should be done. The tomato cut like so, the cucumber like this. At the end of it all we had a feast. Disaster was narrowly averted just before serving when Ayisha, taking a can from the fridge, was about to spoon tuna cat food over the salad. I stopped her just in time, realising later that she would be quite unused to the concept of buying a special can of food to feed a cat.

Sandy and I wanted to ask the plumber to join us. He had just overcharged us for something or other, which he'd turned up to do only after repeated phone calls, but it was seven at night and we couldn't eat while he was watching on hungrily. Ayisha said we were mad.

'Either he gets your money or your food, but not both,' she declared.

I told her that one of the things I liked about Moroccan culture was the way people shared food. Whenever I was travelling by train, people would offer me some of theirs.

'Then they are stupid and are going to starve,' Ayisha said.

'So you include us in that?' I asked.

'No, not you, I am talking about them.'

Sandy and I got our way and the plumber ate with us. He and Ayisha had an animated conversation about how people used to fast during the harvest month leading up to Ramadan, as well as during Ramadan itself. Now adherence to religious principles was becoming less strict and many no longer did this. The plumber still fasted for the additional month, as did Ayisha's mother, although Ayisha and the rest of her family did not.

Ramadan would be upon us in a couple of weeks. As I found it a hardship going without food for more than four waking hours, the prospect of a month-long daytime fast, let alone two months, was inconceivable.

There was excitement in the air in the lead-up to Ramadan. People got their best clothes ready and stockpiled special foods. Apart from not eating in daylight hours, Muslims are not permitted to smoke or have sex during Ramadan. Pregnant or menstruating women are exempt from the dietary restrictions, as are the sick and elderly. All the Moroccans we knew took this commitment to God very seriously, as a form of spiritual purification. So seriously, in fact, that it proved to be a bit of a worry when Ramadan came round.

One day, a week or so into Ramadan, I found the plasterer's assistant rolling around on the floor and groaning from severe stomach cramps. Playing my usual Florence Nightingale role, I produced a couple of Imodium tablets. No, I was told by half a dozen workers, the assistant could not let anything pass his lips. Why not? I wanted to know. After all, the sick are exempt from Ramadan.

'He can only take medicine if he is dying,' Mustapha said.

Listening to the groans of the agonised young man for the next few hours, it certainly sounded like he was.

The *zellijis* – tile craftsmen – had started with a bang, completing their work in the bathroom in just two days. This was no mean feat, since there were four hundred handmade tiles to the square metre. We were thrilled, and thought the completion of all our tile work was imminent, but the next day they didn't turn up. When we rang the chief *zelliji* we were told, very sorry, but they had other work.

It seemed that contractors the world over had developed a technique to ensure the maximum amount of work for them and an equivalent amount of frustration for those employing them. Their credo seemed to be, start a job and get the clients committed, then nick off to start another job, until you have several on the go at once and everyone screaming to get theirs finished. Whoever screams the loudest or pays the most wins.

I threatened and cajoled, saying we were on a deadline and they were holding up the plasterer, and the chief *zelliji* promised they'd work Friday, the usual day off. They did indeed come on Friday – for two whole hours, before discovering there wasn't enough white tiles to finish. So they went home for couscous with Mama instead.

It had always been intriguing to see who of our contractors would show up each morning, although our usual team was reliable and hardworking. But one day, our big gentle worker from the Sahara didn't arrive, and I was astounded to be told he was on strike and wanted a pay rise. He figured he worked harder than anyone else and he wanted more money to compensate. Anxious to avoid a wages breakout, I said we'd find someone else, but

Mustapha claimed he was such a good worker he was worth the extra money. I relented and got Si Mohamed to call him and agree to an increase, provided he kept it confidential. He returned to work within half an hour, smug satisfaction all over his face.

When the team spirit was needed, though, it was there in spades. After a day with eighteen workers on site, we had a dinner party for five friends to prepare. The entire work crew remained after knock-off time and transformed the place in ten minutes flat, sweeping up bucketloads of dirt, washing the floor of the downstairs salon, shifting the big dining table and chairs into it, even though the plaster on the walls was still wet. Sandy and I did the cooking in a corner of the courtyard, wearing mountaineer's headlamps to see what we were doing.

David, who had just returned from several weeks in the States, walked around the house open-mouthed, ogling the changes. 'You guys have managed to do more in three months than I have in five years,' he marvelled. We pointed out that he had a full-time job, while we were dedicated solely to the project and had a minimum of twelve people a day working on it.

But it was true that in the previous month the pace of work had picked up. Now the *halka* was finished, the catwalk was under way, the kitchen had been plastered, along with the downstairs salon – which also had stained glass in the windows – the trenches in the courtyard had been filled, and the bathroom was close to complete. I had designed some furniture and had it made – a dining table and chairs, and a wrought-iron sofa with vibrant red cushions. Things were on the up. With the salon doors shut, we could almost pretend the place was finished.

This was pure illusion, of course, since opening the door revealed a courtyard that looked like a missile had landed there. But we were feeling optimistic that we might even get the bulk of the work done before we had to return to Australia.

My trip to Granada had renewed my interest in the work of the refugees from Al Andalus, so one morning I set off for the Andalusian quarter, with no particular plan in mind other than to follow the main street and see where it led.

The character of this quarter is subtly different, its buildings more compact and not as tall as those in other parts of Fez. Close to the top of the hill on Derb Yasmina, a heavy gateway frames a *masharabbia* screen door. This is the entry to the Medersa Sahrij, one of three religious colleges surrounding the thirteenth-century Jamaa Andalous Mosque, where theology students live and study in tiny rooms. Inside the gate is a most exquisite place.

I found myself standing in a long rectangular courtyard bound by beautifully decorated walls. In the middle was a large pool of a clear aqua colour, also rectangular, fed by a low circular fountain at one end. I was immediately transported back to the Alhambra, except that instead of being among hordes of tourists, I was alone. The work in both places was similar, and as they had been constructed during the fourteenth century, perhaps some artisans had worked on both. Every square centimetre of the Medersa Sahrij's walls had the same elaborate carved plaster as the Alhambra, and the balustrades were made of detailed *masharabbia*. The columns along the two longest sides of the courtyard were covered with *zellij*.

What a place to wake up in, I thought, savouring the harmonious atmosphere created by the perfect symmetry and exceptional workmanship.

Back out on the street, I noticed an unassuming, two-storey building with a small entrance opposite Jamaa Andalous Mosque.

'It's the only working caravanserai left in Fez,' said a middle-aged, djellaba-clad man nearby who saw me looking. 'People who come in from the country can stay for five dirhams a night, and keep their horses downstairs.'

Five dirhams a night? Not a bad deal.

The shops in this section of Fez were only single-storey, and in among them I found another intriguing doorway. A workman was just leaving, and peering over his shoulder I saw that inside was a *zaouia*, a place of worship for Sufis, with tombs of Sufi saints covered in *zellij*. I had a soft spot for Sufis, not just from the ceremonies I'd been to, but because the brotherhoods supported the mentally ill and the disabled, among others.

Sandy departed again, this time for two weeks in Australia to launch his latest book. The plumber, being a fundamentalist, was disconcerted to learn that Sandy wasn't around, and yet he had a habit of arriving at odd times, often when everyone else had gone. Once, I came downstairs and interrupted him changing into his work clothes, and before you could say *humdillilah*, he had changed back again and was out the door to the mosque.

But he'd done enough work by now that I was close to being able to have a real shower. I could almost taste it. I planned to stand under that hot water for a long time, after which I might saunter into the kitchen and turn on my chrome tap, admiring the way the hot water mixed so delightfully with the cold. The way it swirled as it went down the plughole, and then actually went somewhere, instead of creating a muddy puddle.

Eventually, though, my impatience got the better of me, and after yet another unsatisfactory reminder phone call to the plumber, I took matters into my own hands. With Si Mohamed and a little help from Noureddine, I got the shower working in half an hour.

Getting the hot water going was a bit more tricky, as the gas cylinder had to be installed upstairs, above the shower, in the section of the house where the catwalk was being rebuilt. Noureddine managed to winch the cylinder up, balancing precariously on the scaffolding. We fiddled around with the gas but were missing a vital connecting hose. It was getting late and we were all tired, so decided to leave it until the next day.

Perhaps sensing he was being usurped, the plumber turned up the following morning and completed what we had started, stopping every so often to complain about our sloppy workmanship.

After all the workers had gone, I undressed for my inaugural shower. The weather had turned chilly and I shivered in the evening air. No matter, the months of waiting were over.

I would soon be standing under copious hot water delivered straight from my own shower. I turned on the tap and waited. And waited. And waited. What could possibly be wrong now? Was it the plumber's revenge for all my evil thoughts about him?

There was nothing for it but to get dressed again and find my way upstairs in the dark, through the precarious reconstruction of the catwalk to the gas bottle. There I discovered that some fuckwit – no, I'll rephrase that – some caring, dedicated member of my staff had switched the cylinder off in consideration of my safety.

Back in the shower, I was relived to hear a *whoosh* as the gas flame caught alight overhead, followed by hot water bursting out of the shower head and over my grateful body. I stood letting the warmth soak into the very core of my being, washing away not only grit but the nagging tension I'd been holding in the pit of my stomach.

The kitchen fittings were going in at last. Noureddine's moment of truth arrived. He cast a grin back over his shoulder at me as seven men in the courtyard hoisted aloft the massive kitchen cupboard he had painstakingly created.

'*Bismillah*,' Mustapha said, and they all echoed the word of thanks to Allah.

As they carried it through to the kitchen, they started to sing a song of dedication in deep, rich voices. Meanwhile Fatima and Halima gave ululating cries. I felt a surge of love for these people who had become such an integral part of my life.

I raced upstairs to keep an eye on proceedings through the *halka*. When they lifted the cupboard over the benchtop, avoiding the tap with difficulty, they found it wouldn't fit against the wall – Noureddine had forgotten to remove several bits of overhanging wood at one end. While the men pushed the other end through the window and held it in place, Noureddine sawed pieces off. Mustapha muttered, 'Noureddine would forget to eat his own breakfast.'

The cabinet was still about a centimetre too big, so someone fetched a chisel and started hacking into the layer of *haarsh* on the wall. Wads of dirt flew about the place, and I tried to ignore the heavy boots clomping on my new benchtop. Finally the cupboard was pushed approximately into place, but then there was another problem. Noureddine had also forgotten that the two kitchen benches differed in size by some ten centimetres, meaning the sections of the cupboard designed to fit neatly over them weren't in the right place. We couldn't simply cut a bit out of the centre to compensate, because an extractor fan had to fit between the two sections.

Then Si Mohamed pointed out that if we cut five centimetres off one side of the cupboard and moved it along, it would be in the right place. I was momentarily flummoxed, but he was right, it would work. Not perfectly but well enough. I knew Noureddine would be glad that the master carpenter, Abdul Rahim, wasn't there to see his stuff-up.

It wasn't Noureddine's greatest day. Walking home after work that afternoon, he was surrounded by three youths who pulled out knives. One of them hit Noureddine on the head with a knife handle, then they robbed him of everything he was carrying — mobile phone, money, and a few tools he'd borrowed.

Distraught but conscious, Noureddine rushed back to Riad Zany and shouted to the men still working. Within a minute, six of them were pouring out of the door with sticks in their hands, ready to go and do serious damage to the men who'd attacked him. Perhaps fortunately, they did not find them.

I was surprised. I had thought muggings in the Medina a rarity, but several of the workers had stories about brothers or friends being robbed in this way. It was less common for foreigners; thieves generally seemed aware that, tourism being such big business, the police would increase their efforts to find them.

We did have one significant theft, however. One afternoon after Sandy had returned from Australia, two men from a lighting shop brought a selection of brass lanterns to the riad so that we could see them *in situ*. We purchased several, and after they left I noticed our iPod was missing. I'd had it on earlier, and one of the men had been left alone in the salon for a few minutes while we checked out a light in the kitchen. There didn't seem to be another explanation, so I went to the shop with Si Mohamed to heavy the man into returning it. Naturally he denied all knowledge, which meant hours at the police station filling in forms.

It wasn't the loss of the iPod I regretted so much as our music collection. Not having a television, this was our one form of entertainment. The silly thing was that the thief hadn't taken the recharger cord, and in Morocco these weren't easy to come by. It had probably been sold on the Steps of a Thousand

Thieves for a hundred dirhams, played for a few hours and then discarded when it ran out of juice.

Then there were the more subtle attempts at filching from us. The district guardian turned up with a load of old bricks he offered to sell me for half a dirham each, instead of the usual one and a half dirhams for new ones. They were handmade and naturally I was interested. There were seven hundred bricks, he assured us. I asked my guys to count them and there were protests from both sides. It would all be fine. Didn't I trust him? In a word, no. When the men finished counting, there were only four hundred bricks.

'Someone must have taken the rest of them last night,' said the guardian.

Of course they had. Because I had him on the back foot, he agreed to give me a discount. I was learning to bargain like a Moroccan.

friend of Ayisha's had recently given birth, and a few days
before Ramadan there was a celebration, to which Sandy
and I were invited. But going along wasn't just a matter of putting on
a clean blouse and a smear of lipstick; Ayisha told me it was
necessary to hire a traditional Moroccan outfit.

She took me to one of the little shops near the Attarine,
where elaborate dresses hung in a rainbow of colours. They
fulfilled every girl's mediaeval-princess fantasy – shiny fabrics,
floor-length, tapered sleeves, gorgeous embroidery. All that was
missing was the wimple.

Ayisha's choice was easy – vibrant pink. I dithered over mine,
and finally chose a deep purple number with a gold underskirt.
Part of the deal was a wide belt with a cord tie at the back, which
held the waist as firmly as a corset and accentuated the breasts.

Ayisha prevailed upon Sandy to come as well.

'But won't it be all women?' he asked.

'Oh no, there will be lots of men,' she promised, telling him to wear his traditional djellaba and yellow babouches.

The evening of the party, Ayisha arrived at the riad dressed in a black leather miniskirt, high leather boots, and a tight pink T-shirt that just covered her elbows. I was taken aback; it was most unlike her usually modest Medina attire. She looked as if she were off to a London nightclub rather than a traditional Moroccan celebration. We would carry our hired glad rags and change at the party, but as we were leaving I realised that my choice of shoes was woefully inadequate. I had only a pair of flat walking sandals, running shoes, and my plastic *hammam* slip-ons.

Ayisha was disbelieving that I didn't have a pair of high heels, showing me the terribly elegant pair of black stilettos she planned to wear. Just in time, I remembered the soft leather embroidered slippers the neighbour had given me, and decided they would do.

Sandy, already wearing his djellaba, felt as though he were off to a fancy-dress party.

A taxi dropped us at a new block of flats on the outskirts of the Ville Nouvelle. Upstairs we were ushered into a tiny apartment whose every available space was taken up with women and babies.

'Where are the men?' Sandy asked.

'They are somewhere else,' Ayisha said. 'Don't worry.'

But he was worried. A woman beside him flopped out a large breast and began to breastfeed, something that was never usually done in mixed company. It seemed that, Sandy being a foreigner, the normal rules didn't apply. I was worried too, because all the women were wearing housedresses and nighties and looked decidedly unglamorous. Were Ayisha and I going to be the only ones in elaborate dresses?

But I needn't have worried. When, after a bit of chitchat, Ayisha declared it was time to change, the women shrugged off their clothes – despite Sandy's presence – and pulled out similarly ornate outfits. Ayisha took a call on her mobile as I struggled into my dress. Another woman helped me put on the belt, pulling the strings at the back tighter than Scarlett O'Hara's.

We were taken up a further flight of stairs and ushered into a large room ringed with brocade-covered banquettes. There was no one in it, apart from three teenage boys mucking around with a makeshift sound system. Moments later it burst into life. The music was electronic Arab pop, cranked up so far past distortion point it hurt my ears. How was I going to stand several hours of this?

Ayisha, quite prepared to make her own party, began dancing around the room in a hip-gyrating, smouldering kind of way. With a lack of available men to make eye contact with, she focused on Sandy and me alternately. She may not have ever actually seduced anyone, but it didn't stop her practising.

A short time later, Ayisha's mother appeared and sat close by, engaged in a cosy chat with her neighbour. During a break in the music Ayisha told me they were discussing her latest prospective husband – a young man from Casablanca on whom Ayisha wasn't keen. Watching the provocative way she danced, it was no wonder her parents were keen to marry her off as quickly as possible. She was an explosion waiting to happen.

Slowly other guests drifted in, until there were about thirty women and fifteen children. But no men. Sandy, meanwhile, had been sitting next to me making funny observations about the guests. When a young woman in a long blue dress also began to dance in an overtly sexual way on the other side of the room, Sandy and I exchanged glances. She shook her hips and shimmied her shoulders, then tied a scarf over her bottom and began to wiggle frantically. The way she swung her long hair from side to side would have made a 1960s go-go dancer envious.

Encouraged by this display, women young and old began to rise from the banquettes and join in. The little girls copied their older sisters, and soon the entire room was a gyrating mass of women, bumping pelvises, holding hands, openly flirting with one another. There was a palpable sense of sexuality in the air. Sandy's eyes were on sticks. So, I thought, this is how Moroccan women party when they get together.

Ayisha's aunt, a plump, jolly woman with a friendly face, got me up to dance with her. Unfamiliar with the music and the rhythms, I tried my best to wiggle along, but it felt odd and unnatural, and my efforts were more like someone having a seizure than a sexy temptress.

It wasn't until around one in the morning, just before the food was served, that Ayisha decided it was time for Sandy to join the men. He was hustled to an apartment in a building across the road, where a group of men had been chanting Koranic verses since sunset. Eventually they blessed a teapot, and Sandy felt hopeful of some liquid refreshment at least. But when the contents were poured it turned out to be room-temperature milk. He had missed dinner, as the men had naturally been served before the women. After another hour or so, they were each given a single date to eat. Well, it was almost Ramadan.

There was more chanting, then the lights in the apartment went out. Was it a power failure? No, the tempo of the chanting increased and the men stood up and started to throw their heads around in a similar fashion to the women. Then the lights came back on, and everyone sat down and listened to a long sermon, musing philosophically about where Allah could be found in the electric current. Riveting stuff.

The austerity of the men's party was remarkable, compared to the hedonism of the women's. The only ritualised thing the women did was go up and admire the baby at some point in the evening, and slip the mother some money. This was shortly before the food arrived – platters of lamb and prune tagine, and whole chickens with olives. After dinner, when the tables were cleared away, the dancing began again.

The particularly uninhibited dancer in blue came over and took my hand, pulling me onto the dancefloor. Someone passed over the scarf of honour, which was tied around my bottom, and I tried my best to emulate the pelvic thrusting and shaking that my partner was doing so effortlessly. The other women started to clap, and for a few brief embarrassing moments, my bottom became the centre of attention.

Ayisha danced over and began trying to draw my attention away from the girl in blue.

'Not like that,' she shouted into my ear. 'Like this.' She demonstrated the elusive wiggle. After a few minutes of my trying to emulate her, she announced, 'That's enough,' in a peevish voice, took my hand and led me to the other side of the room.

I felt like a small child, and could almost hear her thinking, She's not your friend, she's mine. I hadn't been the object of female jealousy since school.

By this time, it was after two a.m. and I wanted to rescue Sandy and go home. But Ayisha was having a great old time and was in no mood to leave.

'Not yet,' she protested. 'Not until after the pastries.' But once the cakes had been distributed, she continued to dance, ignoring the increasingly glazed looks of her elders. Eventually her mother got up and started to the door. Ayisha followed to bring her back but I hemmed her in from the rear. With an impeccable sense of timing, Sandy met us on the stairs. Ayisha was outnumbered and we were finally able to make our escape.

Once Ramadan began, the streets were eerily quiet in the mornings as the city slept off the binge-eating of the night before. The skies were noticeably clearer without the daytime smoke from cooking fires. After a few days, people assumed a hollow-eyed, haunted look from inadequate sleep and no food or water during daylight. After a couple of weeks, they looked like faded photocopies of their former selves.

Night had turned into day. At sunset, after the cannon had sounded, came the call to prayer. This was the signal for everyone to hoe into a Ramadan breakfast of dates, milk and *harira* soup – lamb or vegetable stock with tomato paste, capsicum, chick peas, lentils, rice, small pasta and coriander. There were side dishes of hardboiled eggs with salt and cumin, and pancake-like breads. To finish, there were deep-fried pastries soaked in sugar syrup, and fruit juice and mint tea.

Between ten p.m. and midnight the 'proper' meal of the day was eaten – either a tagine or couscous or *b'stilla*, a kind of flaky pastry filled with minced pigeon or chicken and topped with, strangely, icing sugar. After this came sleep, then you woke at four a.m. for another breakfast, before going to sleep again. It was customary for street singers to walk the alleys in the small hours, waking people in a lyrical fashion to remind them to eat before dawn.

One sunset we had Ramadan breakfast with Si Mohamed and his family, and watched in admiration as he sat at the table, confronted by food, then as the call to prayer sounded disappeared to the mosque instead of slaking his thirst and hunger. His sisters restrained themselves until their mother had finished her prayers and Si Mohamed returned. We all said a thankful '*Bismillah*' before tucking in.

It was hard for the manual labourers during Ramadan. At the riad, the pace of work slacked off noticeably in the afternoons, so we cut the working day by an hour. And tempers frayed more easily. A neighbour appeared one morning complaining, not for the first time, that we had knocked plaster off his wall with our banging. I had already agreed to pay to repair the cracks, but now he said it was a matter of urgency that this was done before *Eid al-Fitr*, the feast at the end of Ramadan, which has as much significance in Muslim countries as Christmas does in Western ones. I told him I would send our plasterer around to fix the damage the following Saturday. As the plasterer was working flat out to finish our place before Sandy's daughter Yvonne and her family arrived, this was some sacrifice.

A few moments later, I was surprised to hear raised voices from upstairs. I took the stairs two at a time to find this same neighbour surrounded by our angry workmen. It turned out that instead of just taking my word, the neighbour had found it necessary to go up and declare to the plasterer, 'Suzanna says you have to come and work at my place on Saturday.'

As I hadn't yet spoken to the plasterer about this, he was understandably put out and an argument ensued, during which, a shocked Si Mohamed told me later, the neighbour used Allah's name as a swear word. All the workers were outraged and downed tools. The plasterer flat-out refused to work for the neighbour at all. I attempted to mediate a situation that was threatening to take on the dimensions of the Danish cartoon crisis.

I bundled the neighbour downstairs and said that if our plasterer would not work for him then he needed to find another one, but I would pay only as much as ours quoted for the job. The neighbour scurried off to find one, and for hours afterwards I could hear the indignant tones of the men upstairs as they discussed the failings of the neighbour. Who needed *Neighbours* or *The Sopranos* when you had a real-life soap opera?

For months now, whenever people asked when the house would be finished, I'd been saying three weeks. Now here we were with just weeks left and there were some things we clearly couldn't manage this time round – taking the roof off the *massreiya* and repairing the carved and painted ceiling, for one. This was a job that couldn't be rushed, and we figured that since it had been there for several centuries, it could hang on until the following year.

Inevitably, as we got closer to the end, events conspired against us. Part of the ceiling in the downstairs toilet collapsed while Mustapha was repairing it. Fortunately he wasn't standing under it at the time, but fixing it required digging up some of the stairs and was a major job.

At no point did getting the plumber to finish his work become any easier. He continued to show up at strange times, when Si Mohamed was not around to translate and the shops were shut, so he was unable to buy the parts he needed. He seemed incapable of thinking ahead, so I pinned him down and got him to make a list of everything he needed to complete the job and asked him to buy them. He refused, saying Si Mohamed should do this, which meant that half of the items were not the right ones and had to be exchanged – more wasted time.

When he'd fiddled with one of the toilets for the third or

fourth time, unable to fix it, I told him to finish the rest of the work first.

'But where are the parts for the hand basin?' he asked me.

'I presume they were on the list that *you* made, so you must now have them.'

He wandered off. A while later, we were thrilled to discover that the hand basin was connected. No more running across the courtyard to wash our hands. We stood and watched with satisfaction as the water swirled down the plughole.

Doing the dishes later that evening, I pulled the plug out to empty the sink and there was a gush of water from the cupboard beneath, soaking my feet and the floor. Opening the cupboard door, I saw that the parts connecting the sink to the pipe had been removed. Well, that explained why our hand basin now worked.

At least the catwalk was almost complete. Our relationship with the temperamental master carpenter Abdul Rahim might have begun badly, but our respect for him had grown. We'd been surprised when he turned up to do the catwalk himself, instead of sending an apprentice, and were even more amazed when he stuck with it, turning up day after day and doing an excellent job. We found out that his previous employers, French people, had sacked him, which was why we had the privilege of his services.

As he had only given us a quote for the structure, I asked how much he wanted to do the decorative facing, a job of a couple of days, and he gave me a price equivalent to more than half the total cost of the catwalk, which I declined. When the work was very nearly finished I looked up to see him putting a straight piece of wood along the top of the catwalk roof. I had been told that when you use traditional, handmade green tiles, as we planned to, a special piece of scalloped wood must be put in place to contain them. But when I asked Abdul Rahim about this he grew irritated, saying if we wanted it done the traditional way we'd have to pay extra.

'It's a piece of decoration,' he explained.

'So do you put it on after the tiles are in place?' I asked.

'No, before.'

'Then how can it be decoration? That's structural.'

Other people, including Mustapha, confirmed that this piece of wood was an essential element in a traditional roof, but Abdul Rahim would not install it without additional payment. As we had settled on a price for a finished job, we said we would pay him the agreed amount minus what it would cost for Noureddine to complete it. I considered simply paying what he asked for the sake of keeping the peace, but it annoyed me that he kept shifting the boundaries and trying to extract more money. We were already paying him, in Moroccan terms, a small fortune. Enough was enough.

When I explained my position he was furious, shouting that he had intended to ask for even more money and would not take less. He refused to accept the final cash payment I was holding out to him and stormed off. A few days later, he rang and said he wanted his money, and funnily enough, accepted what was offered.

To finish the catwalk we needed to buy handmade roofing tiles, the same green tiles that grace the roofs of traditional buildings all over Fez. They are made, as they have always been, at Ein Knockby on the outskirts of the city, where the same kilns are used to fire the famous Fez blue pottery and the multi-hued tiles. From any vantage point in Fez, clouds of black smoke rising from these kilns are visible.

Mustapha, Si Mohamed and I went to make the purchase together, taking a petit taxi then walking the last half-kilometre. The air was redolent with the sweet smell of the olive pits used for firing. The tile factory was in a courtyard, surrounded by charmless concrete workshops. Mustapha had been coming here since the current owner's father ran the place. The man's grandfather and great-grandfather had run the place before him.

Next to a pile of tiles destined for the roof of the Karaouiyine Mosque, we located the smaller ones we needed – half-flowerpot shapes, with the lower half glazed deep green and the top left unglazed. While waiting for our thousand-odd order to be loaded, I watched a craftsman making tiles.

He took a rectangle of clay, flipped it onto a wooden form that resembled a trowel, and then wet the clay. He made a few little flares around the edges before placing the tile in a row on the ground to dry in the sun. The whole thing was done with an ease and skill born of years of practice. Around him, the ground was covered in tiles waiting to be put into the kiln.

The artisans of Fez work hard at their trades. The guilds they belong to have survived for hundreds of years, although their numbers decreased after the French arrived, due to the import of mass-produced European goods. These days they're taking even more of a hammering with competition from the Chinese.

But such is the importance of the artisans' guilds that there is still one day a year, in September, when their survival is celebrated. We had given our workers the afternoon off to watch the procession of guilds and Sufi brotherhoods from the Bab Bou Jeloud to the tomb of Moulay Idriss II, one of the most revered shrines in Morocco. Moulay Idriss II is now regarded as a saint who watches over the city he created.

Late in the afternoon, Sandy and I had set off to join them. Thousands of Fassis were lining the streets and taking up every available vantage point on the surrounding rooftops, legs dangling from the parapets. We squeezed into a spot on the step of a shop.

The sound of trumpets heralded the entrance of the silk-makers through the blue gate, followed by the brass makers, metal-workers, shoemakers, blacksmiths and merchants, all accompanied by pipes and drums and cheers from the crowd. When some of the musicians paused to perform a fleet-footed dance, the roars of approval could have been heard from the Ville Nouvelle.

A camel stalked regally down the Tala'a Sghira, followed by a group of men balancing silver *tyafar* on their heads. These tagine-shaped containers contained symbolic offerings, such as small cakes, to be placed at the tomb. Other men carried oversized candles to illuminate the tomb's interior, and behind them came an enormous banner on which Koranic verses were embroidered in silk.

Next came a four-tiered box, encased in red cloth embroidered with verses from the Koran and carried with great ceremony on the heads of specially chosen weavers. This box was known as *al-Kaswa*, and each year a new one was placed on the tomb of Moulay Idriss II. Following it were four men carrying an out-stretched cloth to gather money for Moulay Idriss's descendants. Then a bull entered, running around in confusion, dispersing onlookers in its path. It was the first of four that would be sacrificed that day as part of the ritual.

The procession over, music and dancing extended well into the night. This celebration had continued for more than a millennium, and I hoped it would continue for at least another yet.

⌣

Ayisha had had a bumpy ride since graduating. She'd quit a job as receptionist at a Moroccan-owned guesthouse when they did not pay her. Then she was offered a position with a foreign company in the Ville Nouvelle, but had a crisis of confidence and took so long to get back to them that the job went to someone else.

And yet she was desperate to move out of the tiny room in which her whole family lived. Her father claimed that if she moved out by herself he would never speak to her again – an unmarried Moroccan woman's reputation was sacred – so she was hoping to rent an apartment and take her parents with her.

I'd put the word about in the expat community that I had a talented Moroccan friend looking for a job, and one day I got a call from the owner of a lovely guesthouse who needed a front-of-house person. We arranged an interview for a Friday morning, but when I went to collect Ayisha she hadn't even got up yet.

'Quick, quick,' I told her, and rang the owner of the guesthouse to say we'd been held up waiting for a craftsman at the riad. Ayisha clattered down the stairs a few minutes later, clad in jeans and a casual top.

'Hmm,' I said, looking her up and down. 'You really need something more professional.' I followed her back upstairs and she fished out a black skirt and a smart shirt, and took her time to dress. She seemed to have no sense of urgency and was at once charming and infuriating, stopping to chatter and pass me gifts of walnuts. I had to subdue an urge to drag her out by the arm.

Finally we were on our way. The guesthouse was one of the nicest in Fez, and the owners were friendly. I knew they were seeing several other applicants, but hoped Ayisha's English proficiency and her engaging personality would impress them. I made myself scarce while she was being interviewed.

'How did it go?' I asked as we walked home.

'Good, I think,' she said. 'The only problem is the timing. They need someone to start straightaway and I want to take a couple of weeks off next month, but I don't know which ones yet.'

'Why?' I asked, knowing this was unlikely to suit the owners.

'Because the American guy is coming and I want to spend some time with him.'

This was the married man with a small baby. Ayisha had told me she'd finished with him, but it seemed the affair had started again.

'He is separated,' she assured me, sensing my disapproval.

'Is he still living with his wife?'

'Yes.'

'Well, then he's not separated. Do you think he's running his life around you?'

This man couldn't even commit himself to an arrival date, let alone his existing marriage. It seemed to me that Ayisha was forgoing the prospect of a good job because of a Cinderella fantasy: a man, who already had a wife, would rescue her from her need to look after herself by whisking her off to foreign climes where money was plentiful and people didn't have to work. She'd hardly be the first young woman to squander opportunities by pursuing a fantasy, but I felt irritated and had to remind myself that it was her life. I only hoped that when she fell to reality the ground wouldn't be too hard.

A few days later, Ayisha bounced into the courtyard at the riad, sexy and radiant in hip-hugging jeans and a daring top that revealed a flash of bra strap.

'The American guy is here,' she breathed.

'So you're having a good time. That's great. What happens next?'

'Oh, probably marriage,' she said airily.

'Then he's leaving his wife and child?'

'Not exactly. He says he can't leave his baby, it wouldn't be fair.'

'But then you can't get married.'

'Well, under Islam, I can.'

I was gobsmacked. Ayisha as a second wife? It was hard to imagine. It was even harder to picture her being happy under such an arrangement – or the first wife, for that matter.

'How is it going to work?' I asked. 'Is he going to take you back to his country and set you up in your own house? How often will he be able to be with you?'

She shrugged, annoyed by my questions. It was apparent she hadn't thought that far ahead, and didn't want to. Polygamy might be more honest than the Western practice of having a bit on the side, but I just couldn't see how it was going to suit Ayisha.

'I think you're selling yourself short before you even start,' I said, but she ran off to meet him, feet hardly touching the ground.

She was back a week or so later to say she'd just been to see a lawyer about the marriage. I was impressed. She had more sense than I'd given her credit for.

'He says I can only become a second wife if the first wife agrees,' she said. 'Do you think she will?'

'Would you, if you were in her position?' I asked, and Ayisha hung her head. She already knew the answer, and just wanted it confirmed.

'So what should I do?' she said.

'Well, you really have no control over him leaving his wife or not. But if that's what you want him to do, you need to set a deadline for him to make a decision. After that, you have to get on with the rest of your life.'

It sounded like a hopeless case, but Ayisha was so infatuated it was going to take a while for her to see it. By the time I left Fez, she still hadn't.

Eid al-Fitr was approaching, the celebration that marked the end of Ramadan. In the days beforehand, people wished each other *'Eid Marbrouka'* – 'Eid blessings.'

Adults wore new djellabas and brightly coloured babouches. Little girls ran through the streets in beaded taffeta dresses; babies were dressed in mini-djellabas and Fez hats or crowns. Walking down the freshly swept alleys of the Medina became an exercise in dodging people bearing huge trays of crescent-shaped pastries filled with almond paste, or other sweet treats being brought back from the bakeries. The souks were seething as people stocked up for the feast.

We gave our workers the traditional few days off and a bonus. The night before the feast, I went out to a surprise birthday party for a friend while Sandy stayed home with a stomach that felt as though it didn't belong to him. When I returned at eleven-thirty Noureddine was still hard at work assembling bed bases and ban-quettes for our visitors, who were shortly to descend. He had spent the last few evenings staying back to finish things, and now here he was on the eve of one of the most significant days of the Muslim calendar, long after most people were at home with their families.

On the morning of the feast, Sandy and I got up at six and walked down to one of the old city gates at R'Cif. It was still dark and men were hurrying to the mosque, carrying rolled-up prayer mats. A few beggars were about, hoping the spirit of *Eid* would result in a bit of extra alms-giving. A donkey trotted past by itself, panniers bulging, and turned up a side alley. It knew exactly where it was headed.

Along with thousands of Fassis, we were going to see in the dawn on top of the hill of Feddane L'Ghorba. We joined a river of people dressed in white.

The first pink light was just touching the fort as we began to climb up past the cemetery, seeking footholds among the scree and rocks on the steep path. Near the summit, we could hear chanting, hoarse and guttural yet strangely melodic. It was a sound that echoed other times, other cultures. It could have been from Tibet, Thailand or mediaeval Europe.

Once at the top, Sandy headed for the men's entrance in the wall surrounding the compound, while I went to the women's area. Through a keyhole gateway, more than six thousand people were trying to cram into a space that could comfortably hold only half that number, so I stayed outside the inner sanctum and found a spot against the wall. The women around me were unrolling their prayer mats, performing prostrations, then settling themselves like birds on a perch. As they took off their shoes, I saw that many of them had elaborate hennaed designs on their hands and feet.

With typical Moroccan generosity, the woman next to me insisted I share her prayer mat, and I sat leaning back on the cold stone wall, eyes closed, feeling the warmth of the sun's rays on my face. The women joined in the chanting, until there was a sea of sound on which I drifted, held aloft by the unified voice of thousands in a ritual older than the city itself.

Strolling through the deserted streets of the Andalusian quarter afterwards, it seemed as though we had the city to ourselves — as if the entire population had somehow been body-snatched, leaving just the shell of buildings, like a film set waiting for the director to shout 'Action!' I loved those buildings of Fez, but I loved the people more. Every piece of wood and brick had been carved and crafted, carted and placed by hand.

It was finally time for Rachid Haloui to pass judgement on the house. He hadn't seen it in months, and while it wasn't completely finished, it was close enough.

Before his arrival, we'd run around in a mad frenzy moving piles of building rubbish and shoving junk into cupboards. Perhaps ridiculously, I felt nervous, but it was important to me that he liked what we'd done.

I needn't have worried. As soon as Rachid walked through the door, his eyes lit up. He drank in the newly plastered courtyard, the renewed catwalk, the exquisitely crafted *balka*. In the *hammam* he saw Mustapha's precise wall of traditional bricks and the old spring that was now the basis of the hand basin. He walked around not saying very much, taking in the subtle colours and the simplicity that had been the basis of our approach. Sandy and I followed in his wake like earnest students.

Eventually Rachid said, 'I like the way you have worked in this house. This is very rare. You have respected not only the architecture but the spirit of this place.'

We breathed a sigh of relief.

'Usually,' Rachid went on, 'people want to make their own fantasy, but you have not done that. You have adapted it, of course, but changed it so that anything new fits in with the old. I would like to bring other clients of mine here, to show them how it should be done.'

Sandy and I smiled at one another, thrilled by his approval. 'We would be honoured,' I said.

Later, when we were drinking tea in the downstairs salon, I asked Rachid if he was optimistic about the future of the Fez Medina and he shook his head sadly.

'I have been very anxious for a long time. The Medina grew in harmony until the beginning of the twentieth century, but now acculturation has become de-culturation. The city no longer has the capacity to adapt.'

He went on to talk about the dangers of a superficial approach to 'rescuing' the Medina, by those who did not fully understand what was being saved. 'You cannot rescue something unless first you know what it should be. The other danger is this new fashion of foreigners buying houses. I have nothing against people coming and restoring, but the way they renovate and modify makes me anxious — the Medina is losing its spirit. If they come to invest money, or with an Orientalist, Arabian-nights fantasy, they do not understand it. The Medina is like a street girl. Everybody takes something but they give nothing back. It must not be a question of taste — you must respect what is. Many people come here and say, 'It's very nice,' and then they remove what is nice. People who want to make a fantasy should build it somewhere else. Don't try to do it in the Medina.'

Was there any hope? I wondered.

'At least in the Fez Medina there are fourteen thousand buildings,' Rachid said, 'whereas Marrakesh has only four thousand, which may mean Fez has a greater degree of resilience. Here we have a chance, unique in the world, to live as they did in the fourteenth century. We can make it more comfortable, of course, but we don't have the right to change it fundamentally. It should not become a theme park.

'This city is also about emotion,' he went on. 'Following an old man in a djellaba down an alley, with everything close and dark, wondering where you're going. Then you enter a house, a courtyard, and see something jaw-dropping. I'm fifty-four and I never come to the Medina without discovering something new.'

And then we were packing to leave. Si Mohamed would look after the house until we returned the following year. Departing was such a wrench, we were tempted to rearrange our lives so we could live in Fez full-time. Sandy had resolved to resign from the radio station and come back to Fez for a few months a year until I could join him. But I wasn't ready to give up working yet – there were still things I wanted to do, and one of us needed to earn an income.

We had a special thankyou lunch for the workers, who were in high spirits and toasted the house with mint tea. We'd managed to organise work for our regular team with another expat, but they promised to try to come back to us for the next stage – repairing the *massreiya*. Sandy had set up a website for the plasterer, who was getting a continual stream of work as a result.

The day we left, they phoned us from their new place of work, the mobile being passed around to wish us *trek salaama* — safe travel. They chattered on happily in Darija and I had no idea what they were saying, but I understood the sentiment. Having shared our lives with them for so long, they felt like family. We would miss them all.

Piling our luggage into a taxi at R'Cif we set off to the train station. Near the main street in the Andalusian quarter, the driver was forced to slow to a crawl to avoid a man performing an exuberant dance, arms akimbo, in the middle of the road. He was a tall thin ragged figure, and his face was suffused with the ecstasy of being alive. I had no idea why he was dancing there, but I knew exactly how he felt.

Afterword

When Sandy returned the following February to supervise the remaining work, Mustapha and his team were overjoyed to see him. Each of them gave him a huge hug and kissed his cheeks repeatedly. On their first day back they made a tour of the riad, pointing out to one another things they had done. The pride they felt in the house was evident.

Even Tigger returned, as Peter and Karen were departing once more for Australia. She did not seem in the least fazed to find herself back in her old haunts, and immediately started chasing sparrows.

A local Sufi brotherhood, the Hamadcha, offered to bestow *Baraka*, a blessing, on Riad Zany, and one Sunday night, fifty guests and twelve musicians packed into the courtyard. Most of the guests were Moroccan, and after a couple of hours the ceremony reached the point where many had danced themselves into a light trance. Several of the women progressed to a frenetic deep trance, including Ayisha, who danced so wildly she needed to be restrained by three people. It was as if she truly were possessed by a djinn, and after she collapsed she said she remembered nothing of the experience.

A local restaurateur made food for everyone in our kitchen, and the following morning the house was a chaotic mess. There came an unexpected knock on the door: it was the decapo ladies, Fatima and Halima, insisting that Sandy leave while they cleaned the place.

Although I was sorry not to be at the ceremony, I received a multitude of emails and photos from people who attended, telling me how much I was missed. Even on the other side of the world, I too felt blessed.

With thanks to David, who enabled Riad Zany to become a reality; Simon and Fred for sharing their knowledge of Sufism; Kareem for his research on circumcision rituals; Karima for her translations; Helen for her timely advice; Meredith for her wonderful editing; and to all the Riad Zany team for their time, effort and energy.